An Introduction to String Theory and D-Brane Dynamics

With Problems and Solutions

2nd Edition

An Introduction to
String Theory and D-Brane Dynamics

With Problems and Solutions

2nd Edition

Richard J Szabo
Heriot-Watt University, UK

ICP
Imperial College Press

Published by

Imperial College Press
57 Shelton Street
Covent Garden
London WC2H 9HE

Distributed by

World Scientific Publishing Co. Pte. Ltd.
5 Toh Tuck Link, Singapore 596224
USA office: 27 Warren Street, Suite 401-402, Hackensack, NJ 07601
UK office: 57 Shelton Street, Covent Garden, London WC2H 9HE

British Library Cataloguing-in-Publication Data
A catalogue record for this book is available from the British Library.

AN INTRODUCTION TO STRING THEORY AND D-BRANE DYNAMICS
(2nd Edition)
With Problems and Solutions

ISBN-13 978-1-84816-622-6
ISBN-10 1-84816-622-2

Printed in Singapore.

To my Dad, for a good long read

Preface to Second Edition

The second edition of this book has improved in two main ways. Firstly, a few minor typos which plagued the first edition have been corrected. Secondly, and most notably, an extra chapter at the end has been added with detailed solutions to all exercises that appear in the main part of the text. These exercises are meant both to fill in some gaps in the main text, and also to provide the student reader with the rudimentary computational skills required in the field. The inclusion of this extra chapter of solutions should help make this second edition more complete and self-contained. It was decided to keep the overall style of the rest of the book intact, and hence the remaining chapters still serve as a brief, concise and quick introduction into the basic aspects of string theory and D-brane physics.

The author would like to thank Laurent Chaminade and others at Imperial College Press for the encouragement to produce this second edition. This work was supported in part by grant ST/G000514/1 "String Theory Scotland" from the UK Science and Technology Facilities Council.

Richard J. Szabo
Edinburgh, 2010

Preface to First Edition

These notes comprise an expanded version of the string theory lectures given by the author at the 31st and 32nd British Universities Summer Schools on Theoretical Elementary Particle Physics (BUSSTEPP) which were held, respectively, in Manchester, England in 2001 and in Glasgow, Scotland in 2002, and also at the Pacific Institute for the Mathematical Sciences *Frontiers of Mathematical Physics* Summer School on "Strings, Gravity and Cosmology" which was held in Vancouver, Canada in 2003. The schools were attended mostly by Ph.D. students in theoretical high-energy physics who had just completed their first year of graduate studies. The lectures were thereby appropriately geared for this level. No prior knowledge of string theory was assumed, but a good background in quantum field theory, introductory level particle physics and group theory was expected. An acquaintance with the basic ideas of general relativity was helpful but not absolutely essential. Some familiarity with supersymmetry was also assumed because the supersymmetry lectures preceeded the string theory lectures at the schools, although the full-blown machinery and techniques of supersymmetry were not exploited to any large extent.

The main references for string theory used during the courses were the standard books on the subject [Green, Schwarz and Witten (1987); Polchinski (1998)] and the more recent review article [Johnson (2001)]. The prerequisite supersymmetry lectures can be found in [Figueroa-O'Farrill (2001)]. Further references are cited in the text, but are mostly included for historical reasons and are by no means exhaustive. Complete sets of references may be found in the various cited books and review articles.

The lectures were delivered in the morning and exercises were assigned. These problems are also included in these notes. Many of them are intended to fill in the technical gaps which due to time constraints were not covered in the lectures. Others are intended to give the student a better grasp of some "stringy" topics. This book has expanded on many aspects of string theory that were addressed during the schools, mainly to make the presentation clearer.

There were six one-hour lectures in total. Since string theory is nowadays such a vast and extensive subject, some focus on the subject material was of course required. The lectures differ perhaps from most introductory approaches since the intent was to provide the student not only with the rudiments of perturbative string theory, but also with an introduction to the more recently discovered non-perturbative degrees of freedom known as "D-branes", which in the past few years have revolutionalized the thinking about string theory and have brought the subject to the forefront of modern theoretical particle physics once again. This means that much of the standard introductory presentation was streamlined in order to allow for an introduction to these more current developments. The hope was that the student will have been provided with enough background to feel comfortable in starting to read current research articles, in addition to being exposed to some of the standard computational techniques in the field.

The basic perturbative material was covered in roughly the first three lectures and comprises chapters 1–4. Lecture 4 (chapter 5) then started to rapidly move towards explaining what D-branes are, and at the same time introducing some more novel stringy physics. Lectures 5 and 6 (chapters 6 and 7) then dealt with D-branes in detail, studied their dynamics, and provided a brief account of the gauge theory/string theory correspondence which has been such an active area of research over the past few years. For completeness, an extra chapter has also been added (chapter 8) which deals with the Ramond–Ramond couplings of D-branes and other novel aspects of D-brane dynamics such as the important "branes within branes" phenomenon. This final chapter is somewhat more advanced and is geared at the reader with some familiarity in differential topology and geometry.

The author is grateful to the participants, tutors and lecturers of the schools for their many questions, corrections, suggestions and criticisms which have all gone into the preparation of these lecture notes. He would especially like to thank M. Alford, C. Davies, J. Forshaw, M. Rozali and G. Semenoff for having organised these excellent schools, and for the encouragement to write up these notes. He would also like to thank J. Figueroa-O'Farrill and F. Lizzi for practical comments on the manuscript.

This work was supported in part by an Advanced Fellowship from the Particle Physics and Astronomy Research Council (U.K.).

Richard J. Szabo
Edinburgh, 2003

Contents

Chapter 1

A Brief History of String Theory

To help introduce the topics which follow and their significance in high energy physics, in this chapter we will briefly give a non-technical historical account of the development of string theory to date, focusing on its achievements, failures and prospects. This will also help to motivate the vast interest in string theory within the particle theory community. It will further give an overview of the material which will follow.

In conventional quantum field theory, the fundamental objects are mathematical points in spacetime, modeling the elementary point particles of nature. String theory is a rather radical generalization of quantum field theory whereby the fundamental objects are extended one-dimensional lines or loops (Fig. 1.1). The various elementary particles observed in nature correspond to different vibrational modes of the string. While we cannot see a string (yet) in nature, if we are very far away from it we will be able to see its point-like oscillations, and hence measure the elementary particles that it produces. The main advantage of this description is that while there are many particles, there is only one string. This indicates that strings could serve as a good starting point for a unified field theory of the fundamental interactions.

This is the idea that emerged by the end of the 1960s from several years of intensive studies of dual models of hadron resonances [Veneziano (1968)]. In this setting, string theory attempts to describe the strong nuclear force. The excitement over this formalism arose from the fact that string S-matrix scattering amplitudes agreed with those found in meson scattering experiments at the time. The inclusion of fermions into the model led to the notion of a supersymmetric string, or "superstring" for short [Neveu and Schwarz (1971); Ramond (1971)]. The massive particles sit on "Regge trajectories" in this setting.

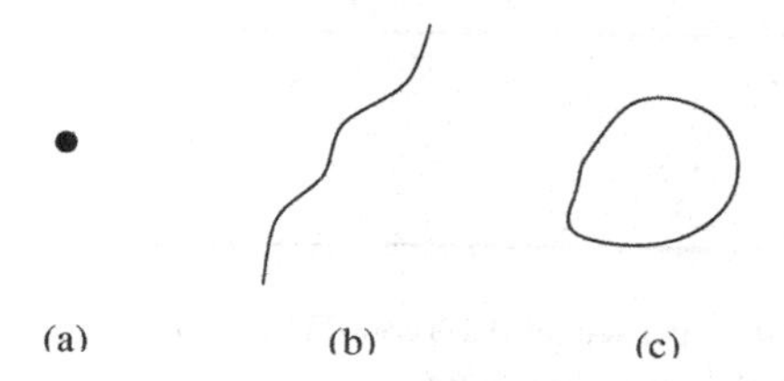

Fig. 1.1 (a) A point particle. (b) An open string. (c) A closed string.

However, around 1973 the interest in string theory quickly began to fade, mainly because quantum chromodynamics became recognized as the correct quantum field theory of the strong interactions. In addition, string theories possessed various undesirable features which made them inappropriate for a theory of hadrons. Among these were the large number of extra spacetime dimensions demanded by string theory, and the existence of massless particles other than the spin 1 gluon in the spectrum of string states.

In 1974 the interest in string theory was revived for another reason [Scherk and Schwarz (1974); Yoneya (1974)]. It was found that, among the massless string states, there is a spin 2 particle that interacts like a graviton. In fact, the only consistent interactions of massless spin 2 particles are gravitational interactions. Thus string theory naturally includes general relativity, and it was thereby proposed as a unified theory of the fundamental forces of nature, including gravity, rather than a theory of hadrons. This situation is in marked contrast to that in ordinary quantum field theory, which does not allow gravity to exist because its scattering amplitudes that involve graviton exchanges are severely plagued by non-renormalizable ultraviolet divergences (Fig. 1.2). On the other hand, string theory is a consistent quantum theory, free from ultraviolet divergences, which necessarily *requires* gravitation for its overall consistency.

With these facts it is possible to estimate the energy or length scale at which strings should be observed in nature. Since string theory is a relativistic quantum theory that includes gravity, it must involve the corresponding three fundamental constants, namely the speed of light c, the reduced Planck constant $\hbar$, and the Newtonian gravitational constant G. These three constants may combined into a constant with dimensions of length. The characteristic length scale of strings may thereby be estimated

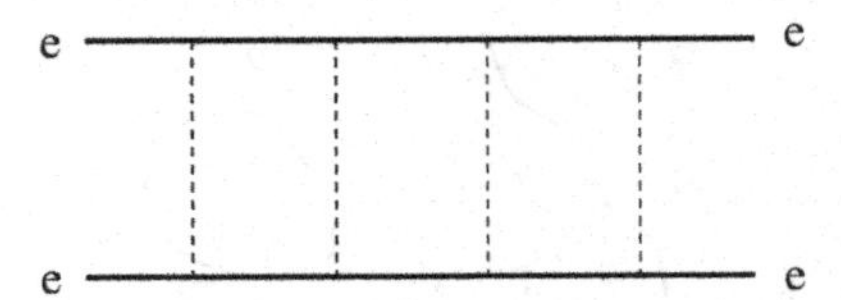

Fig. 1.2 A non-renormalizable ultraviolet divergent Feynman diagram in quantum gravity. The dashed lines depict graviton exchanges.

by the *Planck length* of quantum gravity:

$$\ell_{\rm P} = \left(\frac{\hbar G}{c^3}\right)^{3/2} = 1.6 \times 10^{-33}\ \text{cm}\ . \tag{1.1}$$

This is to be compared with the typical size of hadrons, which is of the order of 10^{-13} cm. The corresponding energy scale is known as the *Planck mass*:

$$m_{\rm P} = \left(\frac{\hbar c}{G}\right)^{1/2} = 1.2 \times 10^{19}\ \text{GeV}/c^2\ . \tag{1.2}$$

These scales indicate the reasons why strings have not been observed in nature thus far. Present day particle accelerators run at energies $\ll m_{\rm P}c^2$ and thus cannot resolve distances as short as the Planck length. At such energies, strings look like point particles, because at very large distance scales compared to the Planck length all one can observe is the string's center of mass motion, which is point-like. Thus at these present day scales, strings are accurately described by quantum field theory.

For many of the subsequent years superstring theory began showing great promise as a unified quantum theory of all the fundamental forces including gravity. Some of the general features which were discovered are:

- General relativity gets modified at very short distances/high energies (below the Planck scale), but at ordinary distances and energies it is present in string theory in exactly the same form as Einstein's theory.

- "Standard model type" Yang–Mills gauge theories arise very naturally in string theory. However, the reasons why the gauge group $SU(3) \times SU(2) \times U(1)$ of the standard model should be singled out is not yet fully understood.
- String theory predicts supersymmetry, because its mathematical consistency depends crucially on it. This is a generic feature of string theory that has not yet been discovered experimentally.

This was the situation for some years, and again the interest in string theory within the high energy physics community began to fade. Different versions of superstring theory existed, but none of them resembled very closely the structure of the standard model.

Things took a sharp turn in 1985 with the birth of what is known as the "first superstring revolution". The dramatic achievement at this time was the realization of how to cancel certain mathematical inconsistencies in quantum string theory. This is known as Green–Schwarz anomaly cancellation [Green and Schwarz (1984)] and its main consequence is that it leaves us with only five consistent superstring theories, each living in ten spacetime dimensions. These five theories are called Type I, Type IIA, Type IIB, $SO(32)$ heterotic, and $E_8 \times E_8$ heterotic. The terminology will be explained later on. For now, we simply note the supersymmetric Yang–Mills gauge groups that arise in these theories. The Type I theories have gauge group $SO(32)$, both Type II theories have $U(1)$, and the heterotic theories have gauge groups as in their names. Of particular phenomenological interest was the $E_8 \times E_8$ heterotic string, because from it one could construct grand unified field theories starting from the exceptional gauge group E_6.

The spacetime dimensionality problem is reconciled through the notion of "compactification". Putting six of the spatial directions on a "small" six-dimensional compact space, smaller than the resolution of the most powerful microscope, makes the 9+1 dimensional spacetime look 3+1 dimensional, as in our observable world. The six dimensional manifolds are restricted by string dynamics to be "Calabi–Yau spaces" [Candelas *et al* (1985)]. These compactifications have tantalizingly similar features to the standard model. However, no complete quantitative agreement has been found yet between the two theories, such as the masses of the various elementary particles. This reason, and others, once again led to the demise of string theory towards the end of the 1980s. Furthermore, at that stage one only understood how to formulate superstring theories in terms of divergent perturbation series analogous to quantum field theory. Like in quantum chromodynamics, it is unlikely that a realistic vacuum can be accurately analysed within perturbation theory. Without a good understanding of nonperturbative effects (such as the analogs of QCD instantons), superstring theory cannot give explicit, quantitative predictions for a grand unified model.

This was the state of affairs until around 1995 when the "second superstring revolution" set in. For the first time, it was understood how to go beyond the perturbation expansion of string theory via "dualities" which probe nonperturbative features of string theory [Font *et al* (1990); Hull and Townsend (1995); Kachru and Vafa (1995); Schwarz (1995); Sen (1994)]. The three major implications of these discoveries were:

- *Dualities relate all five superstring theories in ten dimensions to one another.*

The different theories are just perturbative expansions of a unique underlying theory $\mathcal{U}$ about five different, consistent quantum vacua [Schwarz (1996); Schwarz (1997)]. Thus there is a completely unique theory of nature, whose equation of motion admits many vacua. This is of course a most desirable property of a unified theory.

- *The theory $\mathcal{U}$ also has a solution called "M-Theory" which lives in 11 spacetime dimensions [Duff (1996); Townsend (1995); Witten (1995)].*

The low-energy limit of M-Theory is 11-dimensional supergravity [Cremmer, Julia and Scheck (1978)]. All five superstring theories can be thought of as originating from M-Theory [Duff (1996); Schwarz (1997)] (see Fig. 1.3). The underlying theory $\mathcal{U}$ is depicted in Fig. 1.4.

- *In addition to the fundamental strings, the theory $\mathcal{U}$ admits a variety of extended nonperturbative excitations called "p-branes", where p is the number of spatial extensions of the objects [Horowitz and Strominger (1991)].*

Especially important in this regard are the "Dirichlet p-branes" [Dai, Leigh and Polchinski (1989); Hořava (1989); Polchinski (1995)], or "D-branes" for short, which are p-dimensional soliton-like hyperplanes in spacetime whose quantum dynamics are governed by the theory of *open strings* whose ends are constrained to move on them (Fig. 1.5).

We will not attempt any description of the theory $\mathcal{U}$, which at present is not very well understood. Rather, we wish to focus on the remarkable

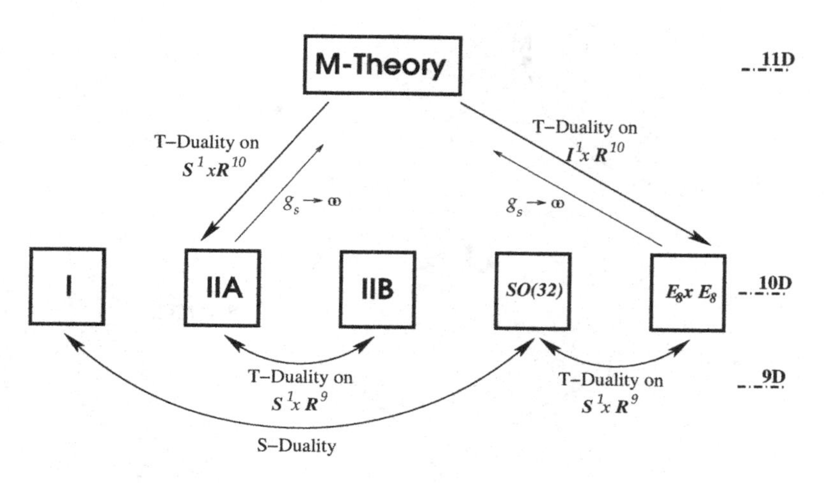

Fig. 1.3 The various duality transformations that relate the superstring theories in nine and ten dimensions. T-Duality inverts the radius R of the circle $\mathbf{S}^1$, or the length of the finite interval $\mathbf{I}^1$, along which a single direction of the spacetime is compactified, i.e. $R \mapsto \ell_{\mathrm{P}}^2/R$. S-duality inverts the (dimensionless) string coupling constant g_s, $g_s \mapsto 1/g_s$, and is the analog of electric-magnetic duality (or strong-weak coupling duality) in four-dimensional gauge theories. M-Theory originates as the strong coupling limit of either the Type IIA or $E_8 \times E_8$ heterotic string theories.

impact in high-energy physics that the discovery of D-branes has provided. Amongst other things, they have led to:

- Explicit realizations of nonperturbative string dualities [Polchinski (1995)]. For example, an elementary closed string state in Theory A (which is perturbative because its amplitudes are functions of the string coupling g_s) gets mapped under an S-duality transformation to a D-brane state in the dual Theory B (which depends on $1/g_s$ and is therefore nonperturbative).
- A microscopic explanation of black hole entropy and the rate of emission of thermal (Hawking) radiation for black holes in string theory [Callan and Maldacena (1996); Strominger and Vafa (1996)].
- The gauge theory/gravity (or AdS/CFT) correspondence [Aharony *et al* (2000); Maldacena (1998)]. D-branes carry gauge fields, while on the other hand they admit a dual description as solutions of the classical equations of motion of string theory and supergravity. Demanding that these two descriptions be equivalent implies, for

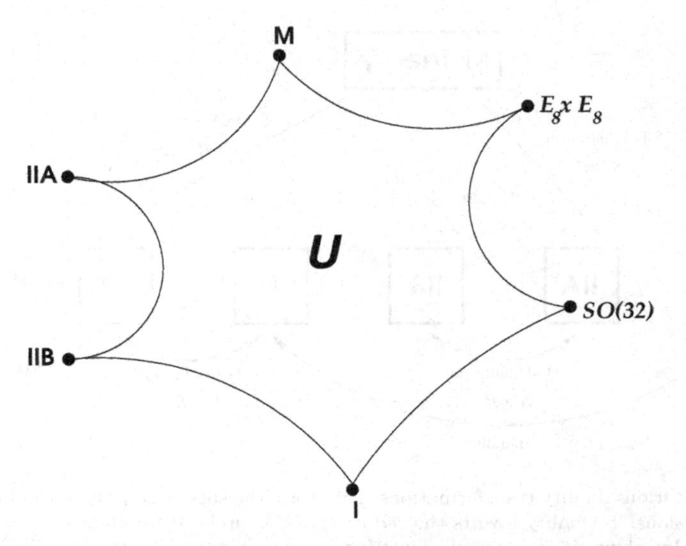

Fig. 1.4 The space $\mathcal{U}$ of quantum string vacua. At each node a weakly-coupled string description is possible.

some special cases, that string theory is equivalent to a gauge field theory. This is an explicit realization of the old ideas that Yang–Mills theory may be represented as some sort of string theory.

- Probes of short-distances in spacetime [Douglas *et al* (1997)], where quantum gravitational fluctuations become important and classical general relativity breaks down.
- Large radius compactifications, whereby extra compact dimensions of size $\gg (\mathrm{TeV})^{-1}$ occur [Antoniadis *et al* (1998); Arkani-Hamed, Dimopoulos and Dvali (1998)]. This is the distance scale probed in present-day accelerator experiments, which has led to the hope that the extra dimensions required by string theory may actually be observable.
- Brane world scenarios, in which we model our world as a D-brane [Randall and Sundrum (1999a); Randall and Sundrum (1999b)]. This may be used to explain why gravity couples so weakly to matter, i.e. why the effective Planck mass in our 3+1-dimensional world is so large, and hence gives a potential explanation of the hierarchy problem $m_{\rm P} \gg m_{\rm weak}$.

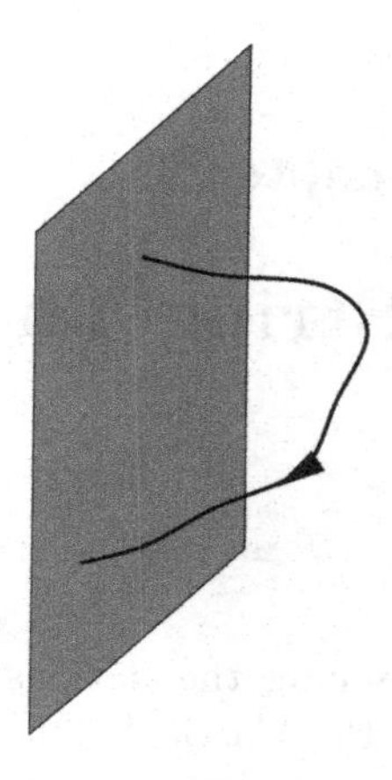

Fig. 1.5 A fundamental open string (wavy line) starting and ending (with Dirichlet boundary conditions) on a Dp-brane (shaded region) which is a $p+1$-dimensional hyperplane in spacetime.

In what follows we will give the necessary background into the description of D-brane dynamics which leads to these exciting developments in theoretical high-energy physics.

Chapter 2

Classical String Theory

In this chapter we will start making the notions of the previous chapter quantitative. We will treat the classical dynamics of strings, before moving on in the next chapter to the quantum theory. Usually, at least at introductory levels, one is introduced to quantum field theory by using "second quantization" which is based on field operators that create or destroy quanta. Here we will describe string dynamics in "first quantization" instead, which is a sum-over-histories or Feynman path integral approach closely tied to perturbation theory. To familiarize ourselves with this geometrical description, we will start by explaining how it can be used to reproduce the well-known dynamics of a massive, relativistic point particle. This will easily introduce the technique that will readily generalize to the case of extended objects such as strings and D-branes. We will then describe the bosonic string within this approach and analyse its classical equations of motion. For the remainder of this book we will work in natural units whereby $c = \hbar = G = 1$.

2.1 The Relativistic Particle

Consider a particle which propagates in d-dimensional "target spacetime" $\mathbb{R}^{1,d-1}$ with coordinates

$$(t, \vec{x}\,) = x^\mu = (x^0, x^1, \dots, x^{d-1}) \; . \tag{2.1}$$

It sweeps out a path $x^\mu(\tau)$ in spacetime, called a "worldline" of the particle, which is parametrized by a proper time coordinate $\tau \in \mathbb{R}$ (Fig. 2.1). The infinitesimal, Lorentz-invariant path length swept out by the particle is

$$\mathrm{d}l = (-\mathrm{d}s^2)^{1/2} = (-\eta_{\mu\nu}\; \mathrm{d}x^\mu\; \mathrm{d}x^\nu)^{1/2} \equiv (-\mathrm{d}x^\mu\; \mathrm{d}x_\mu)^{1/2} \; , \tag{2.2}$$

where l is the proper-time of the particle and

$$(\eta_{\mu\nu}) = \begin{pmatrix} -1 & & & 0 \\ & 1 & & \\ & & \ddots & \\ 0 & & & 1 \end{pmatrix} \tag{2.3}$$

is the flat Minkowski metric. The action for a particle of mass m is then given by the total length of the trajectory swept out by the particle in spacetime:

$$\boxed{S[x] = -m \int \mathrm{d}l(\tau) = -m \int \mathrm{d}\tau \, \sqrt{-\dot{x}^\mu \, \dot{x}_\mu}} \tag{2.4}$$

where

$$\dot{x}^\mu \equiv \frac{\mathrm{d}x^\mu}{\mathrm{d}\tau} \, . \tag{2.5}$$

The minima of this action determine the trajectories of the particle with the smallest path length, and therefore the solutions to the classical equations of motion are the geodesics of the free particle in spacetime.

Exercise 2.1. *Show that the Euler–Lagrange equations resulting from the action (2.4) give the usual equations of relativistic particle kinematics:*

$$\dot{p}^\nu = 0 \, , \quad p^\nu = \frac{m \, \dot{x}^\nu}{\sqrt{-\dot{x}^\mu \, \dot{x}_\mu}} \, .$$

2.1.1 *Reparametrization Invariance*

The Einstein constraint $p^2 \equiv p^\mu \, p_\mu = -m^2$ on the classical trajectories of the particle (Exercise 2.1) is related to the fact that the action (2.4) is invariant under arbitrary, local reparametrizations of the worldline, i.e.

$$\tau \longmapsto \tau(\tau') \, , \quad \frac{\mathrm{d}\tau}{\mathrm{d}\tau'} > 0 \, . \tag{2.6}$$

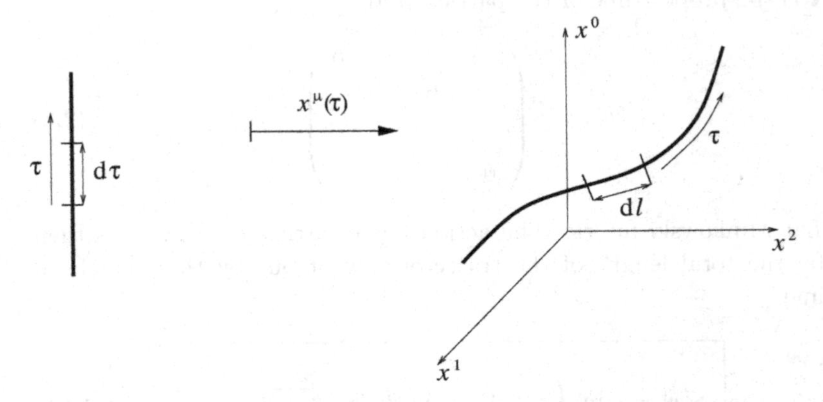

Fig. 2.1 The embedding $\tau \mapsto x^\mu(\tau)$ of a particle trajectory into d-dimensional spacetime. As τ increases the particle propagates along its one-dimensional worldline in the target space.

This is a kind of local worldline "gauge invariance" which means that the form of $S[x]$ is unchanged under such a coordinate change, since

$$\mathrm{d}\tau \ \sqrt{-\dot{x}^\mu(\tau)\dot{x}_\mu(\tau)} = \mathrm{d}\tau' \ \sqrt{-\dot{x}^\mu(\tau')\dot{x}_\mu(\tau')} \ . \tag{2.7}$$

It is a one-dimensional version of the usual four-dimensional general coordinate invariance in general relativity, in the sense that it corresponds to a worldline diffeomorphism symmetry of the theory. An application of the standard Noether procedure to the 0+1-dimensional field theory (2.4) leads to a conserved Noether current whose continuity equation is precisely the constraint $p^2 = -m^2$. It tells us how to eliminate one of the p^μ's, and in the quantum theory it becomes the requirement that physical states and observables must be gauge invariant. It enables us to select a suitable gauge. Let us look at a couple of simple examples, as this point will be crucial for our later generalizations.

2.1.2 *Examples*

Example 2.1. The static gauge choice corresponds to taking

$$x^0 = \tau \equiv t \tag{2.8}$$

in which case the action assumes the simple form

$$S[x] = -m \int \mathrm{d}t \, \sqrt{1 - v^2} \tag{2.9}$$

where

$$\vec{v} = \frac{\mathrm{d}\vec{x}}{\mathrm{d}t} \tag{2.10}$$

is the velocity of the particle. The equations of motion in this gauge take the standard form of those for a free, massive relativistic particle:

$$\frac{\mathrm{d}\vec{p}}{\mathrm{d}t} = \vec{0} \ , \quad \vec{p} = \frac{m\,\vec{v}}{\sqrt{1-v^2}} \ . \tag{2.11}$$

Example 2.2. An even simpler gauge choice, known as the Galilean gauge, results from selecting

$$\dot{x}^\mu \, \dot{x}_\mu = -1 \ . \tag{2.12}$$

The momentum of the particle in this gauge is given by the non-relativistic form (cf. Exercise 2.1)

$$p^\mu = m \, \dot{x}^\mu \ , \tag{2.13}$$

and the equations of motion are therefore

$$\ddot{x}^\mu = 0 \tag{2.14}$$

whose solutions are given by the Galilean trajectories

$$x^\mu(\tau) = x^\mu(0) + p^\mu \, \tau \ . \tag{2.15}$$

2.2 The Bosonic String

In the previous section we analysed a point particle, which is a zero-dimensional object described by a one-dimensional worldline in spacetime. We can easily generalize this construction to a *string*, which is a one-dimensional object described by a two-dimensional "worldsheet" that the string sweeps out as it moves in time with coordinates

$$(\xi^0, \xi^1) = (\tau, \sigma) \ . \tag{2.16}$$

Here $0 \leq \sigma \leq \pi$ is the spatial coordinate along the string, while $\tau \in \mathbb{R}$ describes its propagation in time. The string's evolution in time is described by functions $x^\mu(\tau, \sigma)$, $\mu = 0, 1, \ldots, d-1$ giving the shape of its worldsheet in the target spacetime (Fig. 2.2). The "induced metric" h_{ab} on the string worldsheet corresponding to its embedding into spacetime is given by the "pullback" of the flat Minkowski metric $\eta_{\mu\nu}$ to the surface,

$$h_{ab} = \eta_{\mu\nu}\, \partial_a x^\mu\, \partial_b x^\nu \ , \tag{2.17}$$

where[1]

$$\partial_a \equiv \frac{\partial}{\partial \xi^a} \ , \quad a = 0, 1 \ . \tag{2.18}$$

An elementary calculation shows that the invariant, infinitesimal area element on the worldsheet is given by

$$\mathrm{d}A = \sqrt{-\det_{a,b}\,(h_{ab})}\ \mathrm{d}^2\xi \ , \tag{2.19}$$

where the determinant is taken over the indices $a, b = 0, 1$ of the 2×2 symmetric nondegenerate matrix (h_{ab}).

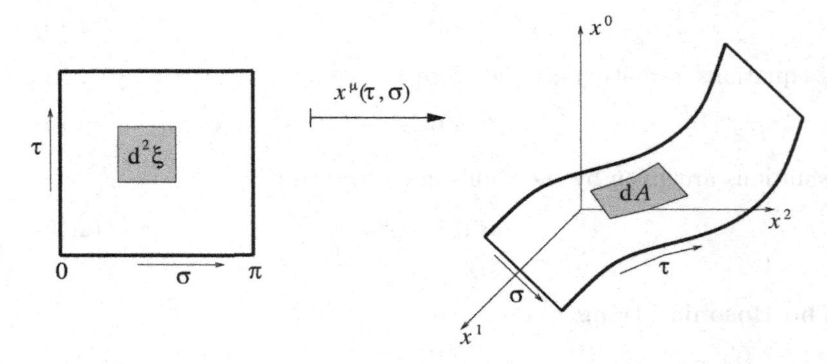

Fig. 2.2 The embedding $(\tau, \sigma) \mapsto x^\mu(\tau, \sigma)$ of a string trajectory into d-dimensional spacetime. As τ increases the string sweeps out its two-dimensional worldsheet in the target space, with σ giving the position along the string.

In analogy to the point particle case, we can now write down an action whose variational law minimizes the total area of the string worldsheet

[1]Notation: Greek letters $\mu, \nu, \ldots$ denote spacetime indices, beginning Latin letters $a, b, \ldots$ denote worldsheet indices, and later Latin letters $i, j, \ldots$ label spatial directions in the target space. Unless otherwise stated, we also adhere to the standard Einstein summation convention for summing over repeated upper and lower indices.

in spacetime:

$$S[x] = -T \int \mathrm{d}A = -T \int \mathrm{d}^2\xi \, \sqrt{-\det_{a,b} \left(\partial_a x^\mu \, \partial_b x_\mu\right)} \ . \tag{2.20}$$

The quantity T has dimensions of mass per unit length and is the *tension* of the string. It is related to the "intrinsic length" ℓ_s of the string by

$$T = \frac{1}{2\pi\alpha'} \ , \quad \alpha' = \ell_s^2 \ . \tag{2.21}$$

The parameter α' is called the "universal Regge slope", because the string vibrational modes all lie on linear parallel Regge trajectories with slope α'. The action (2.20) defines a 1+1-dimensional field theory on the string worldsheet with bosonic fields $x^\mu(\tau,\sigma)$.

Exercise 2.2. *Show that the action (2.20) is reparametrization invariant, i.e. if $\xi \mapsto \xi(\xi')$, then it takes the same form when expressed in terms of the new worldsheet coordinates ξ'.*

Evaluating the determinant explicitly in (2.20) leads to the form

$$S[x] = -T \int \mathrm{d}\tau \ \mathrm{d}\sigma \ \sqrt{\dot{x}^2 \, x'^2 - (\dot{x} \cdot x')^2} \tag{2.22}$$

where

$$\dot{x}^\mu = \frac{\partial x^\mu}{\partial \tau} \ , \quad x'^\mu = \frac{\partial x^\mu}{\partial \sigma} \ . \tag{2.23}$$

This is the form that the original string action appeared in and is known as the "Nambu–Goto action" [Goto (1971); Nambu (1974)]. However, the square root structure of this action is somewhat ackward to work with. It can, however, be eliminated by the fundamental observation that the

Nambu–Goto action is classically equivalent to another action which does not have the square root:

$$
\boxed{\begin{aligned} S[x,\gamma] &= -\frac{T}{2}\int \mathrm{d}^2\xi\ \sqrt{-\gamma}\,\gamma^{ab}\,h_{ab} \\ &= -\frac{T}{2}\int \mathrm{d}^2\xi\ \sqrt{-\gamma}\,\gamma^{ab}\,\partial_a x^\mu\,\partial_b x^\nu\,\eta_{\mu\nu}\ . \end{aligned}} \tag{2.24}
$$

Here the auxilliary rank two symmetric tensor field $\gamma_{ab}(\tau,\sigma)$ has a natural interpretation as a metric on the string worldsheet, and we have defined

$$
\gamma = \det_{a,b}(\gamma_{ab})\ ,\quad \gamma^{ab} = (\gamma^{-1})^{ab}\ . \tag{2.25}
$$

The action (2.24) is called the "Polyakov action" [Polyakov (1981a)].

Exercise 2.3. *Show that the Euler-Lagrange equations obtained by varying the Polyakov action with respect to* γ^{ab} *are*

$$
T_{ab} \equiv \partial_a x^\mu\,\partial_b x_\mu - \frac{1}{2}\gamma_{ab}\,\gamma^{cd}\,\partial_c x^\mu\,\partial_d x_\mu = 0\ ,\quad a,b=0,1\ .
$$

Then show that this equation can be used to eliminate the worldsheet metric γ_{ab} *from the action, and as such recovers the Nambu–Goto action.*

2.2.1 *Worldsheet Symmetries*

The quantity T_{ab} appearing in Exercise 2.3 is the energy–momentum tensor of the 1+1-dimensional worldsheet field theory defined by the action (2.24). The conditions $T_{ab}=0$ are often refered to as "Virasoro constraints" [Virasoro (1970)] and they are equivalent to two local "gauge symmetries" of the Polyakov action, namely the "reparametrization invariance"

$$
(\tau,\sigma) \longmapsto \Big(\tau(\tau',\sigma')\,,\,\sigma(\tau',\sigma')\Big)\ , \tag{2.26}
$$

and the "Weyl invariance" (or "conformal invariance")

$$
\gamma_{ab} \longmapsto \mathrm{e}^{\,2\rho(\tau,\sigma)}\,\gamma_{ab} \tag{2.27}
$$

where $\rho(\tau,\sigma)$ is an arbitrary function on the worldsheet. These two local symmetries of $S[x,\gamma]$ allow us to select a gauge in which the three functions residing in the symmetric 2×2 matrix (γ_{ab}) are expressed in terms of just a *single* function. A particularly convenient choice is the "conformal gauge"

$$(\gamma_{ab}) = \mathrm{e}^{\,\phi(\tau,\sigma)}\,(\eta_{ab}) = \mathrm{e}^{\,\phi(\tau,\sigma)} \begin{pmatrix} -1 & 0 \\ 0 & 1 \end{pmatrix} . \tag{2.28}$$

In this gauge, the metric γ_{ab} is said to be "conformally flat", because it agrees with the Minkowski metric η_{ab} of a flat worldsheet, but only up to the scaling function $\mathrm{e}^{\,\phi}$. Then, at the classical level, the conformal factor $\mathrm{e}^{\,\phi}$ drops out of everything and we are left with the simple gauge-fixed action $S[x,\,\mathrm{e}^{\,\phi}\,\eta]$, i.e. the Polyakov action in the conformal gauge, and the constraints $T_{ab} = 0$:

$$\boxed{\begin{aligned} S[x,\,\mathrm{e}^{\,\phi}\,\eta] &= T \int \mathrm{d}\tau\ \mathrm{d}\sigma\ \left(\dot{x}^2 - x'^2\right) , \\ T_{01} = T_{10} &= \dot{x} \cdot x' = 0 , \\ T_{00} = T_{11} &= \frac{1}{2}\left(\dot{x}^2 + x'^2\right) = 0 . \end{aligned}} \tag{2.29}$$

Note that apart from the constraints, (2.29) defines a *free* (noninteracting) field theory.

2.3 String Equations of Motion

The equations of motion for the bosonic string can be derived by applying the variational principle to the 1+1-dimensional field theory (2.29). Varying the Polyakov action in the conformal gauge with respect to the x^μ gives

$$\delta S[x,\,\mathrm{e}^{\,\phi}\,\eta] = T \int \mathrm{d}\tau\ \mathrm{d}\sigma\ \left(\eta^{ab}\,\partial_a\,\partial_b x_\mu\right)\,\delta x^\mu - T \int \mathrm{d}\tau\ x'_\mu\,\delta x^\mu \Big|_{\sigma=0}^{\sigma=\pi} . \tag{2.30}$$

The first term in (2.30) yields the usual bulk equations of motion which here correspond to the two-dimensional wave equation

$$\left(\frac{\partial^2}{\partial\sigma^2} - \frac{\partial^2}{\partial\tau^2}\right) x^\mu(\tau,\sigma) = 0 . \tag{2.31}$$

The second term comes from the integration by parts required to arrive at the bulk differential equation, which involves a total derivative over the spatial interval $0 \leq \sigma \leq \pi$ of the string. In order that the total variation of the action be zero, these boundary terms must vanish as well. The manner in which we choose them to vanish depends crucially on whether we are dealing with *closed* or *open* strings. The solutions of the classical equations of motion then correspond to solutions of the wave equation (2.31) with the appropriate boundary conditions.

Closed Strings : Here we tie the two ends of the string at $\sigma = 0$ and $\sigma = \pi$ together by imposing periodic boundary conditions on the string embedding fields (Fig. 2.3):

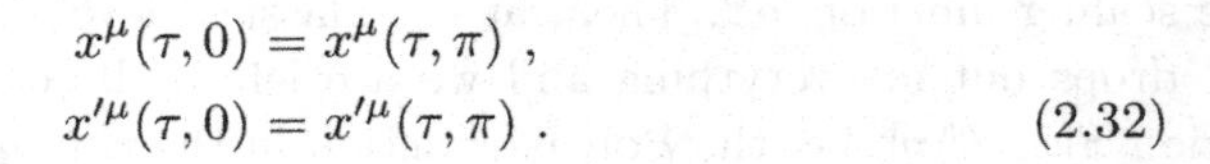

$$
\begin{aligned}
x^\mu(\tau, 0) &= x^\mu(\tau, \pi) \ , \\
x'^\mu(\tau, 0) &= x'^\mu(\tau, \pi) \ .
\end{aligned}
\tag{2.32}
$$

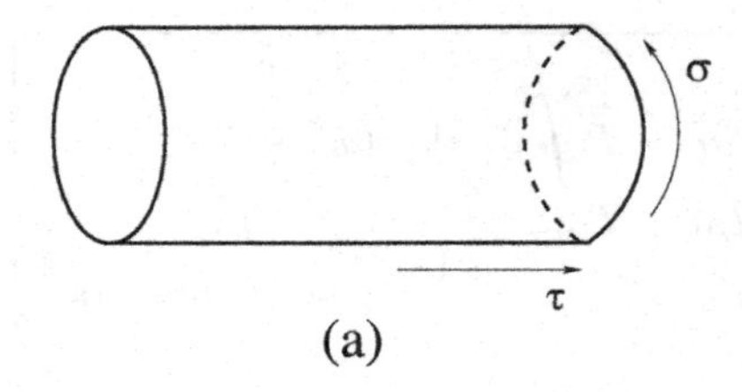

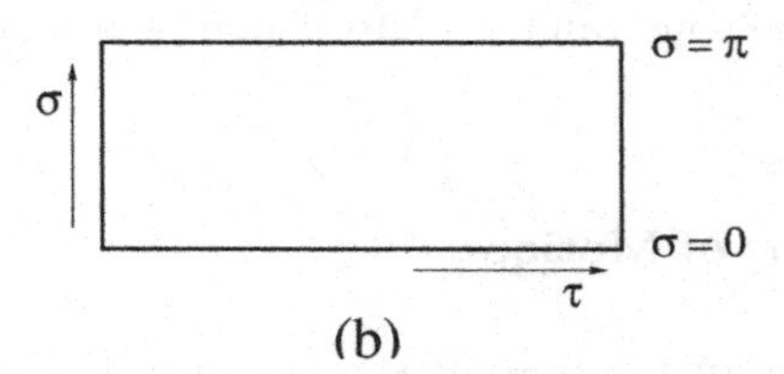

Fig. 2.3 The worldsheet of (a) a closed string is an infinite cylinder $\mathbb{R} \times \mathbf{S}^1$, and of (b) an open string is an infinite strip $\mathbb{R} \times \mathbf{I}^1$.

Open Strings : Here there are two canonical choices of boundary conditions.[2] *Neumann* boundary conditions are defined by

$$x'^\mu(\tau, \sigma)\Big|_{\sigma=0,\pi} = 0 \ . \tag{2.33}$$

In this case the ends of the string can sit anywhere in spacetime. *Dirichlet* boundary conditions, on the other hand, are defined by

$$\dot{x}^\mu(\tau, \sigma)\Big|_{\sigma=0,\pi} = 0 \ . \tag{2.34}$$

[2]These are by no means the only boundary conditions possible, but they are the only ones of direct physical relevance to the topics covered in this book.

Integrating the condition (2.34) over τ specifies a spacetime location on which the string ends, and so Dirichlet boundary conditions are equivalent to fixing the endpoints of the string:

$$\delta x^\mu(\tau,\sigma)\Big|_{\sigma=0,\pi} = 0 \ . \tag{2.35}$$

We will see later on that this spacetime point corresponds to a physical object called a "D-brane". For the time being, however, we shall focus our attention on Neumann boundary conditions for open strings.

2.3.1 *Mode Expansions*

To solve the equations of motion, we write the two-dimensional wave equation (2.31) in terms of worldsheet light-cone coordinates:

$$\partial_+\partial_- x^\mu = 0 \ , \tag{2.36}$$

where

$$\xi^\pm = \tau \pm \sigma \ , \quad \partial_\pm = \frac{\partial}{\partial \xi^\pm} \ . \tag{2.37}$$

The general solution of (2.36) is then the sum of an analytic function of ξ^+ alone, which we will call the "left-moving" solution, and an analytic function of ξ^- alone, which we call the "right-moving" solution, $x^\mu(\tau,\sigma) = x^\mu_{\rm L}(\xi^+) + x^\mu_{\rm R}(\xi^-)$. The precise form of the solutions now depends on the type of boundary conditions.

Closed Strings : The periodic boundary conditions (2.32) accordingly restrict the Taylor series expansions of the analytic functions which solve (2.36), and we arrive at the solution

$$\boxed{\begin{aligned} x^\mu(\tau,\sigma) &= x^\mu_{\rm L}(\xi^+) + x^\mu_{\rm R}(\xi^-) \ , \\ x^\mu_{\rm L}(\xi^+) &= \frac{1}{2}\,x^\mu_0 + \alpha'\,p^\mu_0\,\xi^+ + {\rm i}\sqrt{\frac{\alpha'}{2}}\,\sum_{n\neq 0}\frac{\tilde\alpha^\mu_n}{n}\ {\rm e}^{-2\,{\rm i}\,n\xi^+} \ , \\ x^\mu_{\rm R}(\xi^-) &= \frac{1}{2}\,x^\mu_0 + \alpha'\,p^\mu_0\,\xi^- + {\rm i}\sqrt{\frac{\alpha'}{2}}\,\sum_{n\neq 0}\frac{\alpha^\mu_n}{n}\ {\rm e}^{-2\,{\rm i}\,n\xi^-} \ . \end{aligned}} \tag{2.38}$$

We have appropriately normalized the terms in these Fourier-type series expansions, which we will refer to as "mode expansions", according to physical dimension. Reality of the string embedding function x^μ requires the integration constants x_0^μ and p_0^μ to be real, and

$$(\tilde{\alpha}_n^\mu)^* = \tilde{\alpha}_{-n}^\mu \ , \quad (\alpha_n^\mu)^* = \alpha_{-n}^\mu \ . \tag{2.39}$$

By integrating x^μ and $\dot{x}^\mu$ over $\sigma \in [0, \pi]$ we see that x_0^μ and p_0^μ represent the center of mass position and momentum of the string, respectively. The $\tilde{\alpha}_n^\mu$ and α_n^μ represent the oscillatory modes of the string. The mode expansions (2.38) correspond to those of left and right moving travelling waves circulating around the string in opposite directions.

Open Strings : For open strings, the spatial worldsheet coordinate σ lives on a finite interval rather than a circle. The open string mode expansion may be obtained from that of the closed string through the "doubling trick", which identifies $\sigma \sim -\sigma$ on the circle $\mathbf{S}^1$ and thereby maps it onto the finite interval $[0, \pi]$ (Fig. 2.4). The open string solution to the equations of motion may thereby be obtained from (2.38) by imposing the extra condition $x^\mu(\tau, \sigma) = x^\mu(\tau, -\sigma)$. This is of course still compatible with the wave equation (2.31) and it immediately implies the Neumann boundary conditions (2.33). We therefore find

$$\boxed{x^\mu(\tau, \sigma) = x_0^\mu + 2\alpha' \, p_0^\mu \, \tau + \mathrm{i}\sqrt{2\alpha'} \sum_{n \neq 0} \frac{\alpha_n^\mu}{n} \, \mathrm{e}^{-\mathrm{i}n\tau} \, \cos(n\sigma) \ .} \tag{2.40}$$

The open string mode expansion has a standing wave for its solution, representing the left and right moving sectors reflected into one another by the Neumann boundary condition (2.33).

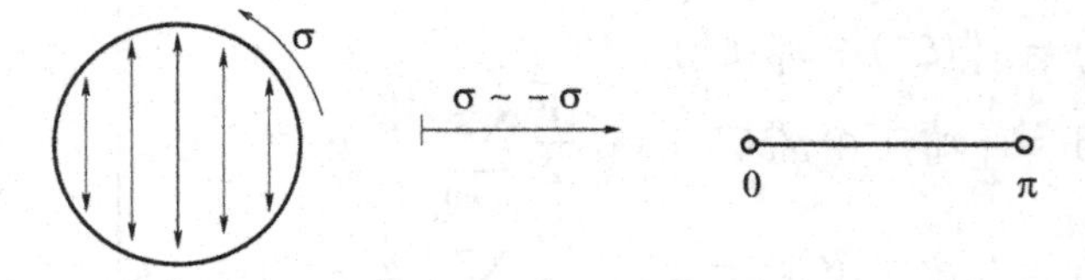

Fig. 2.4 The doubling trick identifies opposite points on the circle and maps it onto a finite interval.

2.3.2 Mass-Shell Constraints

The final ingredients to go into the classical solution are the physical constraints $T_{ab} = 0$. In the light-cone coordinates (2.37), the components T_{-+} and T_{+-} are identically zero, while in the case of the closed string the remaining components are given by

$$T_{++}(\xi^+) = \frac{1}{2}\left(\partial_+ x_{\rm L}\right)^2 = \sum_{n=-\infty}^{\infty} \tilde{L}_n \, {\rm e}^{2\,{\rm i}\,n\xi^+} = 0 \,,$$

$$T_{--}(\xi^-) = \frac{1}{2}\left(\partial_- x_{\rm R}\right)^2 = \sum_{n=-\infty}^{\infty} L_n \, {\rm e}^{2\,{\rm i}\,n\xi^-} = 0 \,, \tag{2.41}$$

where we have defined

$$\tilde{L}_n = \frac{1}{2}\sum_{m=-\infty}^{\infty} \tilde{\alpha}_{n-m}\cdot\tilde{\alpha}_m \,, \quad L_n = \frac{1}{2}\sum_{m=-\infty}^{\infty} \alpha_{n-m}\cdot\alpha_m \tag{2.42}$$

with

$$\tilde{\alpha}_0^\mu = \alpha_0^\mu = \sqrt{\frac{\alpha'}{2}}\, p_0^\mu \,. \tag{2.43}$$

For open strings, we have only the constraint involving untilded quantities, and the definition of the zero modes (2.43) changes to $\alpha_0^\mu = \sqrt{2\alpha'}\, p_0^\mu$.

This gives an infinite number of constraints corresponding to an infinite number of conserved currents of the 1+1-dimensional field theory. They are associated with the residual, local, infinite-dimensional "conformal symmetry" of the theory which preserves the conformal gauge condition (2.28):

$$\xi^+ \longmapsto \xi'^+ = f(\xi^+) \,,$$

$$\xi^- \longmapsto \xi'^- = g(\xi^-) \,, \tag{2.44}$$

where f and g are arbitrary analytic functions. Only the conformal factor ϕ in (2.28) is affected by such coordinate transformations, and so the entire classical theory is invariant under them. They are known as "conformal transformations" and they rescale the induced worldsheet metric while leaving preserved all angles in two-dimensions. This "conformal invariance" of the worldsheet field theory makes it a "conformal field theory" [Belavin, Polyakov and Zamolodchikov (1984); Ginsparg (1990)], and represents one of the most powerful results and techniques of perturbative string theory.

Chapter 3

Quantization of the Bosonic String

We will now proceed to quantize the classical theory of the previous chapter. We will proceed by applying the standard canonical quantization techniques of quantum field theory to the 1+1-dimensional worldsheet field theory. Imposing the mass-shell constraints will thereby lead to the construction of physical states and the appearence of elementary particles in the quantum string spectrum. We will then turn to some more formal aspects of the quantum theory, namely a description of the operators which create particles from the string vacuum, a heuristic explanation of the structure of the string perturbation expansion, and how to incorporate non-abelian gauge degrees of freedom into the spectrum of string states.

3.1 Canonical Quantization

Recall from the previous chapter that the string worldsheet field theory is essentially a free field theory, and hence its quantization is particularly simple. The basic set-up for canonical quantization follows the standard practise of quantum field theory and is left as an exercise.

Exercise 3.1. **(a)** *Starting from the Polyakov action in the conformal gauge, show that the total classical Hamiltonian is given by*

$$H = \begin{cases} \dfrac{1}{2}\displaystyle\sum_{n=-\infty}^{\infty} \alpha_{-n}\cdot\alpha_n = L_0 & \text{for open strings ,} \\ \dfrac{1}{2}\displaystyle\sum_{n=-\infty}^{\infty} (\tilde{\alpha}_{-n}\cdot\tilde{\alpha}_n + \alpha_{-n}\cdot\alpha_n) = \tilde{L}_0 + L_0 & \text{for closed strings .} \end{cases}$$

(b) *Calculate the canonical momentum conjugate to the string embedding field x^μ, and hence show that the oscillator modes have the quantum commutators*

$$
\begin{aligned}
[x_0^\mu\,,\,p_0^\nu] &= \mathrm{i}\,\eta^{\mu\nu}\,, \\
[x_0^\mu\,,x_0^\nu] &= [p_0^\mu\,,p_0^\nu] \;=\; 0\,, \\
[\alpha_m^\mu\,,\,\alpha_n^\nu] &= [\tilde\alpha_m^\mu\,,\,\tilde\alpha_n^\nu] \;=\; m\,\delta_{m+n,0}\,\eta^{\mu\nu}\,, \\
[\alpha_m^\mu\,,\,\tilde\alpha_n^\nu] &= 0\,.
\end{aligned}
$$

From Exercise 3.1 it follows that $(a_m^\mu, a_m^{\mu\,\dagger}) = (\frac{1}{\sqrt{m}}\,\alpha_m^\mu, \frac{1}{\sqrt{m}}\,\alpha_{-m}^\mu)$ define quantum mechanical raising and lowering operators for the simple harmonic oscillator, i.e.

$$
\left[a_m^\mu\,,\,a_m^{\mu\,\dagger}\right] = 1\,. \tag{3.1}
$$

This structure is of course anticipated, because free quantum fields are simply composed of infinitely many harmonic oscillators. The corresponding Hilbert space is therefore a Fock space spanned by products of states $|n\rangle$, $n = 0, 1, 2, \ldots$, which are built on a normalized ground state $|0\rangle$ annihilated by the lowering operators $a = a_m^\mu$:

$$
\begin{aligned}
\langle 0|0\rangle &= 1\,, \\
a|0\rangle &= 0\,, \\
|n\rangle &= \frac{(a^\dagger)^n}{\sqrt{n!}}\,|0\rangle\,.
\end{aligned} \tag{3.2}
$$

By repeated use of the commutation relation (3.1), in the usual way one can arrive at the relations

$$
\begin{aligned}
\langle m|n\rangle &= \delta_{nm}\,, \\
a^\dagger a|n\rangle &= n|n\rangle\,.
\end{aligned} \tag{3.3}
$$

In each sector of the theory, i.e. open, and closed left-moving and right-moving, we get d independent families of such infinite sets of oscillators, one for each spacetime dimension $\mu = 0, 1, \ldots, d-1$. The only subtlety in this case is that, because $\eta^{00} = -1$, the time components are proportional to oscillators with the wrong sign, i.e. $[a_m^0, a_m^{0\,\dagger}] = -1$. Such oscillators are potentially dangerous, because they create states of negative norm which can lead to an inconsistent, non-unitary quantum theory (with negative probabilities and the like). However, as we will see, the Virasoro constraints

$T_{ab} = 0$ eliminate the negative norm states from the physical spectrum of the string.

From Exercise 3.1 it also follows that the zero mode operators x_0^μ and p_0^ν obey the standard Heisenberg commutation relations. They may thereby be represented on the Hilbert space spanned by the usual plane wave basis $|k\rangle = \mathrm{e}^{\,\mathrm{i}\,k\cdot x}$ of eigenstates of p_0^μ. Thus the Hilbert space of the string is built on the states $|k;0\rangle$ of center of mass momentum k^μ with

$$\begin{aligned} p_0^\mu|k;0\rangle &= k^\mu|k;0\rangle \ , \\ \alpha_m^\mu|k;0\rangle &= 0 \end{aligned} \tag{3.4}$$

for $m > 0$ and $\mu = 0, 1, \ldots, d-1$. For closed strings, there is also an independent left-moving copy of the Fock space.

3.1.1 *Normal Ordering*

For the quantum versions of the Virasoro operators defined in (2.42), we use the usual "normal ordering" prescription that places all lowering operators to the right. Then the L_n for $n \neq 0$ are all fine when promoted to quantum operators. However, the Hamiltonian L_0 needs more careful definition, because α_n^μ and α_{-n}^μ do not commute. As a quantum operator we have

$$L_0 = \frac{1}{2}\,\alpha_0^2 + \sum_{n=1}^{\infty} \alpha_{-n}\cdot\alpha_n + \varepsilon_0 \ , \tag{3.5}$$

where

$$\varepsilon_0 = \frac{d-2}{2}\sum_{n=1}^{\infty} n \tag{3.6}$$

is the total zero-point energy of the families of infinite field oscillators (coming from the usual zero point energy $\frac{1}{2}$ of the quantum mechanical harmonic oscillator). The factor of $d-2$ appears, rather than d, because after imposition of the physical constraints there remain only $d-2$ independent polarizations of the string embedding fields x^μ. Explicitly, the constraints can be satisfied by imposing a spacetime light-cone gauge condition $p^+ = 0$ which retains only transverse degrees of freedom.

The quantity ε_0 is simply the Casimir energy arising from the fact that the 1+1-dimensional quantum field theory here is defined on a box, which is an infinite strip for open strings and an infinite cylinder for closed strings. Of course $\varepsilon_0 = \infty$ formally, but, as usual in quantum field theory, it can be regulated to give a finite answer corresponding to the total zero-point energy of all harmonic oscillators in the system. For this, we write it as

$$\varepsilon_0 = \frac{d-2}{2}\,\zeta(-1)\ , \tag{3.7}$$

where

$$\zeta(z) = \sum_{n=1}^{\infty} \frac{1}{n^z} \tag{3.8}$$

for $z \in \mathbb{C}$ is called the "Riemann zeta-function". The function $\zeta(z)$ can be regulated and has a well-defined analytical continuation to a finite function for $\mathrm{Re}(z) \leq 1$. In this paper we will only require its values at $z = 0$ and $z = -1$ [Gradshteyn and Ryzhik (1980)]:

$$\begin{aligned} \zeta(0) &= -\frac{1}{2}\ , \\ \zeta(-1) &= -\frac{1}{12}\ . \end{aligned} \tag{3.9}$$

The vacuum energy is thereby found to be

$$\boxed{\varepsilon_0 = -\frac{d-2}{24}\ .} \tag{3.10}$$

Exercise 3.2. **(a)** *Show that the angular momentum operators of the worldsheet field theory are given by*

$$J^{\mu\nu} = x_0^\mu\,p_0^\nu - x_0^\nu\,p_0^\mu - \mathrm{i}\sum_{n=1}^{\infty}\frac{1}{n}\left(\alpha_{-n}^\mu\,\alpha_n^\nu - \alpha_{-n}^\nu\,\alpha_n^\mu\right)\ .$$

(b) *Use the canonical commutation relations to verify that the Poincaré algebra*

$$\begin{aligned} \left[p_0^\mu\,,\,p_0^\nu\right] &= 0\ , \\ \left[p_0^\mu\,,\,J^{\nu\rho}\right] &= -\mathrm{i}\,\eta^{\mu\nu}\,p_0^\rho + \mathrm{i}\,\eta^{\mu\rho}\,p_0^\nu\ , \\ \left[J^{\mu\nu}\,,\,J^{\rho\lambda}\right] &= -\mathrm{i}\,\eta^{\nu\rho}\,J^{\mu\lambda} + \mathrm{i}\,\eta^{\mu\rho}\,J^{\nu\lambda} + \mathrm{i}\,\eta^{\nu\lambda}\,J^{\mu\rho} - \mathrm{i}\,\eta^{\mu\lambda}\,J^{\nu\rho} \end{aligned}$$

is satisfied.
(c) *Show that for all n,*

$$\left[L_n\,,\,J^{\mu\nu}\right] = 0\ .$$

This will guarantee later on that the string states are Lorentz multiplets.

3.2 The Physical String Spectrum

Our construction of quantum states of the bosonic string will rely heavily on a fundamental result that is at the heart of the conformal symmetry described at the end of section 2.3.

Exercise 3.3. *Use the oscillator algebra to show that the operators L_n generate the infinite-dimensional "Virasoro algebra of central charge $c = d$" [Virasoro (1970)],*

$$[L_n, L_m] = (n-m)\, L_{n+m} + \frac{c}{12}\left(n^3 - n\right)\delta_{n+m,0}\ .$$

The constant term on the right-hand side of these commutation relations is often called the "conformal anomaly", as it represents a quantum breaking of the classical conformal symmetry algebra [Polyakov (1981a)].

We define the "physical states" $|\mathrm{phys}\rangle$ of the full Hilbert space to be those which obey the Virasoro constraints $T_{ab} \equiv 0$:

$$\begin{aligned}(L_0 - a)|\mathrm{phys}\rangle &= 0\ , \quad a \equiv -\varepsilon_0 > 0\ ,\\ L_n|\mathrm{phys}\rangle &= 0\end{aligned} \tag{3.11}$$

for $n > 0$. These constraints are just the analogs of the "Gupta–Bleuler prescription" for imposing mass-shell constraints in quantum electrodynamics. The L_0 constraint in (3.11) is a generalization of the Klein–Gordon equation, as it contains $p_0^2 = -\partial_\mu\, \partial^\mu$ plus oscillator terms. Note that because of the central term in the Virasoro algebra, it is inconsistent to impose these constraints on both L_n and L_{-n}.

3.2.1 *The Open String Spectrum*

We will begin by considering open strings as they are somewhat easier to describe. Mathematically, their spectrum is the same as that of the right-moving sector of closed strings. The closed string spectrum will thereby be straightforward to obtain afterwards. The constraint $L_0 = a$ in (3.11) is then equivalent to the "mass-shell condition"

$$\boxed{m^2 = -p_0^2 = -\frac{1}{2\alpha'}\,\alpha_0^2 = \frac{1}{\alpha'}\left(N - a\right)\ ,} \tag{3.12}$$

where N is the "level number" which is defined to be the oscillator number operator

$$N = \sum_{n=1}^{\infty} \alpha_{-n} \cdot \alpha_n = \sum_{n=1}^{\infty} n \, a_n^\dagger \cdot a_n = 0, 1, 2, \ldots \; , \tag{3.13}$$

and $N_n \equiv a_n^\dagger \cdot a_n = 0, 1, 2, \ldots$ is the usual number operator associated with the oscillator algebra (3.1).

Ground State $N = 0$: The ground state has a unique realization whereby all oscillators are in the Fock vacuum, and is therefore given by $|k; 0\rangle$. The momentum k of this state is constrained by the Virasoro constraints to have mass-squared given by

$$-k^2 = m^2 = -\frac{a}{\alpha'} < 0 \; . \tag{3.14}$$

Since the vector k^μ is space-like, this state therefore describes a "tachyon", i.e. a particle which travels faster than the speed of light. So the bosonic string theory is *not* a consistent quantum theory, because its vacuum has imaginary energy and hence is *unstable.* As in quantum field theory, the presence of a tachyon indicates that one is perturbing around a local maximum of the potential energy, and we are sitting in the wrong vacuum. However, in perturbation theory, which is the framework in which we are implicitly working here, this instability is not visible. Since we will eventually remedy the situation by studying tachyon-free superstring theories, let us just plug along without worrying for now about the tachyonic state.

First Excited Level $N = 1$: The only way to get $N = 1$ is to excite the first oscillator modes once, $\alpha_{-1}^\mu |k; 0\rangle$. We are also free to specify a "polarization vector" ζ_μ for the state. So the most general level 1 state is given by

$$|k; \zeta\rangle = \zeta \cdot \alpha_{-1} |k; 0\rangle \; . \tag{3.15}$$

The Virasoro constraints give the energy

$$m^2 = \frac{1}{\alpha'} \Big(1 - a\Big) \; . \tag{3.16}$$

Furthermore, using the commutation relations and (2.42) we may compute

$$L_1 |k; \zeta\rangle = \sqrt{2\alpha'} \, (k \cdot \alpha_1)(\zeta \cdot \alpha_{-1}) |k; 0\rangle = \sqrt{2\alpha'} \, (k \cdot \zeta) |k; 0\rangle \; , \tag{3.17}$$

and thus the physical state condition $L_1 |k; \zeta\rangle = 0$ implies that the polarization and momentum of the state must obey

$$k \cdot \zeta = 0 \; . \tag{3.18}$$

The cases $a \neq 1$ turn out to be unphysical, as they contain tachyons and ghost states of negative norm. So we shall take $a = 1$, which upon comparison with (3.10) fixes the "bosonic critical dimension of spacetime":

$$\boxed{d = 26\ .} \tag{3.19}$$

The condition (3.19) can be regarded as the requirement of cancellation of the conformal anomaly in the quantum theory [Polyakov (1981a)], obtained by demanding the equivalence of the quantizations in both the light-cone and conformal gauges. The latter approach relies on the introduction of worldsheet Faddeev–Popov ghost fields for the gauge-fixing of the conformal gauge in the quantum theory.

Exercise 3.4. *Consider the "spurious state" defined by* $|\psi\rangle = L_{-1}|k;0\rangle$. *Show that:*

(a) *It can be written in the form*

$$|\psi\rangle = \sqrt{2\alpha'}\ |k;k\rangle\ .$$

(b) *It is orthogonal to any physical state.*

(c)
$$L_1|\psi\rangle = 2\alpha'\ k^2|k;0\rangle\ .$$

The $N = 1$ state $|k;\zeta\rangle$ constructed above with $k^2 = m^2 = 0$ and $k\cdot\zeta = 0$ has $d - 2 = 24$ independent polarization states, as the physical constraints remove two of the initial d vector degrees of freedom. It therefore describes a massless spin 1 (vector) particle with polarization vector ζ_μ, which agrees with what one finds for a massless Maxwell or Yang–Mills field. Indeed, there is a natural way to describe an associated gauge invariance in this picture.

Gauge Invariance : The corresponding spurious state $|\psi\rangle$ is both physical and null, and so we can add it to *any* physical state with no physical consequences. We should therefore impose an equivalence relation

$$|\text{phys}\rangle \sim |\text{phys}\rangle + \lambda\,|\psi\rangle \tag{3.20}$$

where λ is any constant. For the physical state $|\text{phys}\rangle = |k;\zeta\rangle$, Exercise 3.4 (a) implies that (3.20) is the same as an equivalence relation on the polarization vectors,

$$\zeta^\mu \sim \zeta^\mu + \lambda \sqrt{2\alpha'}\, k^\mu \ . \tag{3.21}$$

This is a $U(1)$ gauge symmetry (in momentum space), and so at level $N = 1$ we have obtained the 24 physical states of a photon field $A_\mu(x)$ in 26-dimensional spacetime.

Higher Levels $N \geq 2$: The higher level string states with $N \geq 2$ are all massive and will not be dealt with here. We simply note that there is an infinite tower of them, thereby making the string theory suited to describe all of the elementary particles of nature.

3.2.2 *The Closed String Spectrum*

The case of closed strings is similar to that of open strings. We now have to also incorporate the left-moving sector Fock states. Thus we can easily extend the analysis to the closed string case by simply taking the tensor product of two copies of the open string result and changing the definition of the zero-modes as in (2.43). However, a new condition now arises. Adding and subtracting the two physical state conditions $(L_0 - 1)|\text{phys}\rangle = (\tilde{L}_0 - 1)|\text{phys}\rangle = 0$ yields, respectively, the quantum constraints

$$\begin{aligned} \left(L_0 + \tilde{L}_0 - 2\right)|\text{phys}\rangle &= 0 \ , \\ \left(L_0 - \tilde{L}_0\right)|\text{phys}\rangle &= 0 \ , \end{aligned} \tag{3.22}$$

where we have again fixed the value $a = 1$. The first constraint yields the usual mass-shell relation, since $H = L_0 + \tilde{L}_0 - 2$ is the worldsheet Hamiltonian which generates time translations on the string worldsheet. The second constraint can be understood by noting that the operator $P = L_0 - \tilde{L}_0$ is the worldsheet momentum, and so it generates translations in the string position coordinate σ. This constraint therefore simply reflects the fact that there is no physical significance as to where on the string we are, and hence that the physics is invariant under translations in σ. It amounts to equating the number of right-moving and left-moving oscillator modes. We thereby arrive at the new mass-shell relation

$$\boxed{m^2 = \frac{4}{\alpha'}\left(N - 1\right) \ ,} \tag{3.23}$$

and the additional "level-matching condition"

$$\boxed{N = \tilde{N}} \ . \tag{3.24}$$

Ground State $N = 0$: The ground state is $|k; 0, 0\rangle$ and it has mass-squared

$$m^2 = -\frac{4}{\alpha'} < 0 \ . \tag{3.25}$$

It again represents a spin 0 tachyon, and the closed string vacuum is also unstable.
First Excited Level $N = 1$: The first excited state is generically of the form

$$|k; \zeta\rangle = \zeta_{\mu\nu} \left(\alpha^{\mu}_{-1} |k; 0\rangle \otimes \tilde{\alpha}^{\nu}_{-1} |k; 0\rangle \right) \tag{3.26}$$

and it has mass-squared

$$m^2 = 0 \ . \tag{3.27}$$

The Virasoro constraints in addition give

$$L_1 |k; \zeta\rangle = \tilde{L}_1 |k; \zeta\rangle = 0 \tag{3.28}$$

which are equivalent to

$$k^{\mu} \, \zeta_{\mu\nu} = 0 \ . \tag{3.29}$$

A polarization tensor $\zeta_{\mu\nu}$ obeying (3.29) encodes three distinct spin states according to the decomposition of $\zeta_{\mu\nu}$ into irreducible representations of the spacetime "Little group" $SO(24)$, which classifies massless fields in this case [Figueroa-O'Farrill (2001)]. This is the residual Lorentz symmetry group that remains after the Virasoro constraints have been taken into account, so that the spectrum is exactly what one would get from 24 vectors in each sector which are transversely polarized to the light-cone. From a group theoretical perspective, this decomposition comes from writing down the Clebsch–Gordan decomposition of the tensor product of two vector representations $\mathbf{24}$ of $SO(24)$ into the irreducible symmetric, antisymmetric, and trivial representations,

$$\mathbf{24} \otimes \mathbf{24} = \mathbf{S} \oplus \mathbf{A} \oplus \mathbf{1} \ . \tag{3.30}$$

More concretely, we decompose the rank 2 tensor $\zeta_{\mu\nu}$ according to (3.30) as

$$\begin{aligned}\zeta_{\mu\nu} &= \left[\frac{1}{2}\left(\zeta_{\mu\nu}+\zeta_{\nu\mu}\right)-\frac{1}{25}\,\mathrm{tr}\,(\zeta)\right]+\left[\frac{1}{2}\left(\zeta_{\mu\nu}-\zeta_{\nu\mu}\right)\right]\\ &\quad+\left[\frac{1}{25}\,\eta_{\mu\nu}\,\mathrm{tr}\,(\zeta)\right]\\ &\equiv [g_{\mu\nu}]+[B_{\mu\nu}]+[\eta_{\mu\nu}\,\Phi]\ . \end{aligned} \tag{3.31}$$

The symmetric, traceless tensor $g_{\mu\nu}$ corresponds to the spin 2 "graviton field" and it yields the spacetime metric. The antisymmetric spin 2 tensor $B_{\mu\nu}$ is called the "Neveu–Schwarz B-field", while the scalar field Φ is the spin 0 "dilaton".
Gauge Invariance : The equivalence relations generated by spurious states built from L_{-1} and $\tilde{L}_{-1}$ give the "gauge transformations"

$$\begin{aligned} g_{\mu\nu} &\longmapsto g_{\mu\nu}+\partial_\mu\Lambda_\nu+\partial_\nu\Lambda_\mu\ ,\\ B_{\mu\nu} &\longmapsto B_{\mu\nu}+\partial_\mu\Lambda_\nu-\partial_\nu\Lambda_\mu\ . \end{aligned} \tag{3.32}$$

In this sense the tensor $g_{\mu\nu}$ admits a natural interpretation as a graviton field, as it has the correct diffeomorphism gauge invariance. Its presence accounts for the fact that string theory contains gravity, which is a good approximation at energies $\ll 1/\ell_s$. Its vacuum expectation value $\langle g_{\mu\nu}\rangle$ determines the spacetime geometry. Similarly, the vacuum expectation value of the dilaton Φ determines the "string coupling constant" g_s through

$$g_s = \left\langle\, \mathrm{e}^{\Phi}\,\right\rangle\ . \tag{3.33}$$

This relationship can be derived using vertex operators, which will be discussed in the next section.

The gauge transformation rule for $B_{\mu\nu}$, on the other hand, is a generalization of that for the Maxwell field:

$$A_\mu \longmapsto A_\mu+\partial_\mu\Lambda\ . \tag{3.34}$$

The importance of the B-field resides in the fact that a fundamental string is a source for it, just like a charged particle is a source for an electromagnetic vector potential A_μ through the coupling

$$q\int \mathrm{d}\tau\ \dot{x}^\mu(\tau)\,A_\mu\ . \tag{3.35}$$

In an analogous way, the B-field couples to strings via

$$q \int \mathrm{d}^2\xi \ \epsilon^{ab} \, \partial_a x^\mu \, \partial_b x^\nu \, B_{\mu\nu} \ , \tag{3.36}$$

where ϵ^{ab} is the antisymmetric tensor with $\epsilon^{01} = -\epsilon^{10} = 1$.

3.2.3 *Worldsheet-Spacetime Interplay*

The upshot of the physical string spectrum can be summarized through the interplay between worldsheet and spacetime quantities. At the lowest level of massless states ($N = 1$), open strings correspond to gauge theory while closed strings correspond to gravity. This interplay will be a recurring theme in these notes. The higher levels $N \geq 2$ give an infinite tower of massive particle excitations. Notice that the massless states are picked out in the limit $\alpha' \to 0$ ($\ell_s \to 0$) in which the strings look like point-like objects. We shall refer to this low-energy limit as the "field theory limit" of the string theory.

3.3 Vertex Operators

Any local and unitary quantum field theory has an appropriate operator-state correspondence which allows one to associate quantum fields to quantum states in a one-to-one manner. The states may then be regarded as being created from the vacuum by the quantum fields. We will now describe how to formulate this correspondence in the context of string theory. For this, we map the closed string cylinder and the open string strip to the complex plane and the upper complex half-plane, respectively, first by a Wick rotation $\tau \mapsto \mathrm{i}\,\tau$ to Euclidean worldsheet signature, followed by the coordinate transformation $z = \mathrm{e}^{\tau - \mathrm{i}\,\sigma}$ (Fig. 3.1). The advantage of this coordinate transformation is that it allows us to reinterpret the mode expansions of section 2.3 as Laurent series in the complex plane. In particular, for closed strings the coordinate transformation $\xi^\pm \mapsto z, \overline{z}$ allows us to write (2.38) as

$$\begin{aligned} \partial_z x^\mu_{\mathrm{L}}(z) &= -\mathrm{i}\sqrt{\frac{\alpha'}{2}} \sum_{n=-\infty}^{\infty} \alpha^\mu_n \, z^{-n-1} \ , \\ \partial_{\overline{z}} x^\mu_{\mathrm{R}}(\overline{z}) &= -\mathrm{i}\sqrt{\frac{\alpha'}{2}} \sum_{n=-\infty}^{\infty} \tilde{\alpha}^\mu_n \, \overline{z}^{-n-1} \ . \end{aligned} \tag{3.37}$$

These relations can be inverted by using the Cauchy integral formula to give

$$\alpha^{\mu}_{-n} = \sqrt{\frac{2}{\alpha'}} \oint \frac{\mathrm{d}z}{2\pi}\, z^{-n}\, \partial_z x^{\mu}_{\mathrm{L}}(z) \ ,$$
$$\tilde{\alpha}^{\mu}_{-n} = \sqrt{\frac{2}{\alpha'}} \oint \frac{\mathrm{d}\overline{z}}{2\pi}\, \overline{z}^{-n}\, \partial_{\overline{z}} x^{\mu}_{\mathrm{R}}(\overline{z}) \ , \tag{3.38}$$

where the contour integrations encircle the origin $z = 0$ of the complex plane with counterclockwise orientation and are non-vanishing for $n \geq 0$. It follows that the states built from the α^{μ}_{-n}'s are related to the residues of $\partial_z^n x^{\mu}_{\mathrm{L}}(z)$ at the origin, where $\partial_z^n x^{\mu}_{\mathrm{L}}(0)$ corresponds to an insertion of a point-like operator at $z = 0$. These operators are called "vertex operators". Let us consider some elementary examples of this operator-state correspondence.

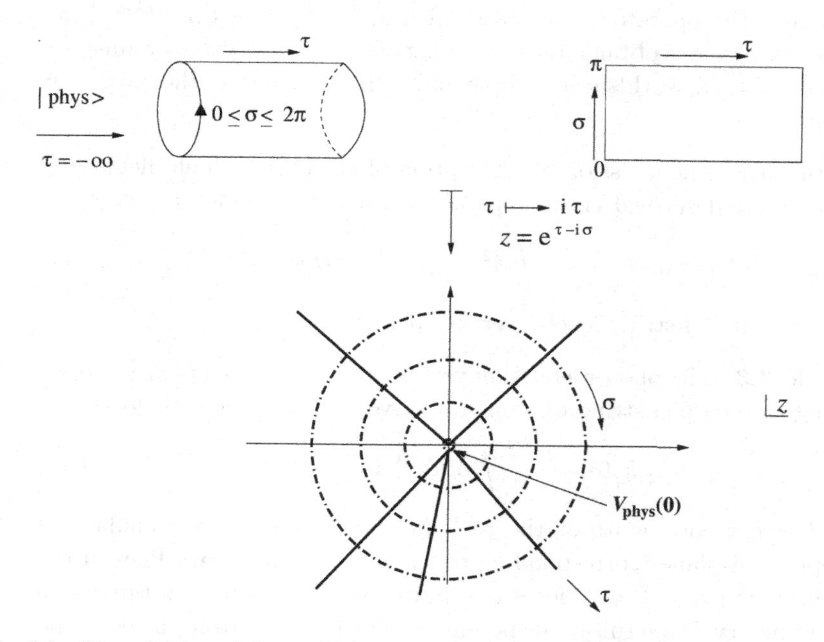

Fig. 3.1 The mapping of the worldsheet cylinder (resp. strip) onto the complex plane (resp. upper complex half-plane). The lines of constant τ form concentric circles about the origin in the complex z-plane, while the lines of constant σ correspond to rays emerging in the radial directions from the origin. An incoming physical state $|\mathrm{phys}\rangle$ in the infinite worldsheet past ($\tau = -\infty$) corresponds to the insertion of a vertex operator $V_{\mathrm{phys}}(z)$ at the origin $z = 0$.

3.3.1 *Examples*

Example 3.1. Geometrically, the closed string tachyon vertex operator is the spacetime Fourier transform of the operator

$$V(x) = \int \mathrm{d}^2 z\ \delta\Big(x - x(z,\overline{z})\Big) \tag{3.39}$$

which pins a string at the spacetime point x. In other words, the correspondence between tachyon ground states $|k;0,0\rangle = \mathrm{e}^{\,\mathrm{i}\,k\cdot x_0}$ and vertex operators is given by

$$|k;0,0\rangle \longleftrightarrow \int \mathrm{d}^2 z\ {}^\circ_\circ\ \mathrm{e}^{\,\mathrm{i}\,k\cdot x(z,\overline{z})}\ {}^\circ_\circ\ , \tag{3.40}$$

where ${}^\circ_\circ \cdot {}^\circ_\circ$ denotes normal ordering of quantum operators. This correspondence agrees with the anticipated behaviour under translations $x^\mu \mapsto x^\mu + c^\mu$ by constant vectors c^μ generated by the target space momentum, under which the operator (and state) pick up a phase factor $\mathrm{e}^{\,\mathrm{i}\,k\cdot c}$. Note that the state here is obtained by averaging over the absorption or emission point on the string worldsheet, as it should be independent of the particular insertion point.

Example 3.2. The emission or absorption of the gravitational fields $g_{\mu\nu}$, $B_{\mu\nu}$ and Φ are described via the operator-state correspondence

$$\zeta_{\mu\nu}\,\alpha_{-1}^\mu\,\tilde{\alpha}_{-1}^\nu|k;0,0\rangle \longleftrightarrow \int \mathrm{d}^2 z\ \zeta_{\mu\nu}\ {}^\circ_\circ\ \partial_z x^\mu\,\partial_{\overline{z}} x^\nu\ \ \mathrm{e}^{\,\mathrm{i}\,k\cdot x}\ {}^\circ_\circ\ , \tag{3.41}$$

defining the closed string level-1 vertex operator.

Example 3.3. The photon-emission vertex operator is the operator corresponding to the open string photon state given by the correspondence

$$\zeta_\mu\,\alpha_{-1}^\mu|k;0\rangle \longleftrightarrow \int \mathrm{d}l\ \zeta_\mu\ {}^\circ_\circ\ \partial_\| x^\mu\ \ \mathrm{e}^{\,\mathrm{i}\,k\cdot x}\ {}^\circ_\circ\ , \tag{3.42}$$

where l is the coordinate of the real line representing the boundary of the upper half-plane (corresponding to the $\sigma = 0,\pi$ boundary lines of the worldsheet strip), and $\partial_\|$ denotes the derivative in the direction tangential to the boundary. From this correspondence, which follows from the doubling trick $z \sim \overline{z}$ of section 2.3, it is evident that the photon is associated with the endpoints of the open string. This is a quantum version of the classical property that the string endpoints move at the speed of light, which can be easily deduced from examining the worldsheet canonical momentum of the open string theory.

3.4 String Perturbation Theory

We will now heuristically describe the structure of the string perturbation expansion, with emphasis on how it differs from that of ordinary quantum field theory. In the latter case, perturbation theory calculations are carried out by computing Feynman diagrams, which are webs of worldlines of point particles. The particles interact at a well-defined point in spacetime where straight lines, whose amplitudes are given by their Feynman propagators, intersect at vertices (Fig. 3.2). A scattering amplitude is then calculated by drawing the corresponding Feynman diagrams, and multiplying together all the propagators and the coupling constants at each vertex.

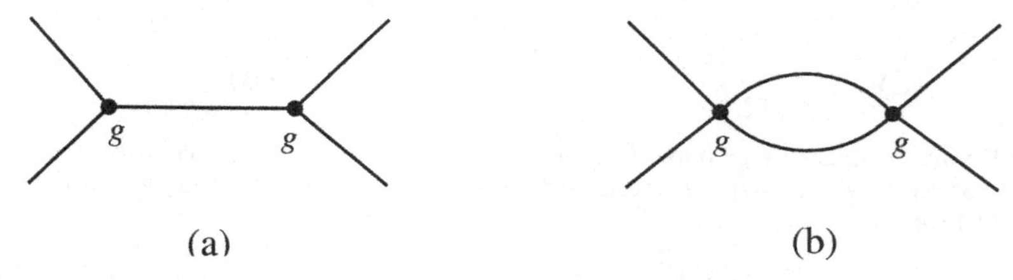

Fig. 3.2 Feynman graphs for four-particle (a) tree-level and (b) one-loop scattering amplitudes in quantum field theory. The lines denote propagators which correspond to the worldlines of particles in spacetime. Each vertex has a coupling constant g associated to it and represents a singular worldline junction responsible for the ultraviolet divergences in loop amplitudes.

In string theory, the situation is similar, except that the Feynman diagrams are *smooth* two-dimensional surfaces representing string worldsheets. The propagation of a string is represented by a tube (Fig. 3.3). It turns out that after including the contributions from the infinite tower of massive particles in the string spectrum, the non-renormalizable divergences in quantum gravity loop amplitudes completely cancel each other out. The particularly significant feature here is that string interactions are "smoother" than the interactions of point particles, because the worldsheets are generically smooth. The ultraviolet divergences of quantum field theory, which are rendered finite in loop amplitudes by strings, can be traced back to the fact that the interactions of quantum field theory are associated with worldline junctions at specific spacetime points. But because the string worldsheet is smooth (with *no* singular points), string theory scattering amplitudes have *no* ultraviolet divergences. A profound consequence of this smoothness property is that in string theory the structure of interactions is completely determined by the *free* worldsheet field theory, and there are no arbitrary interactions to be chosen. The interaction is a consequence of worldsheet topology (the "handles"), rather than of local singularities.

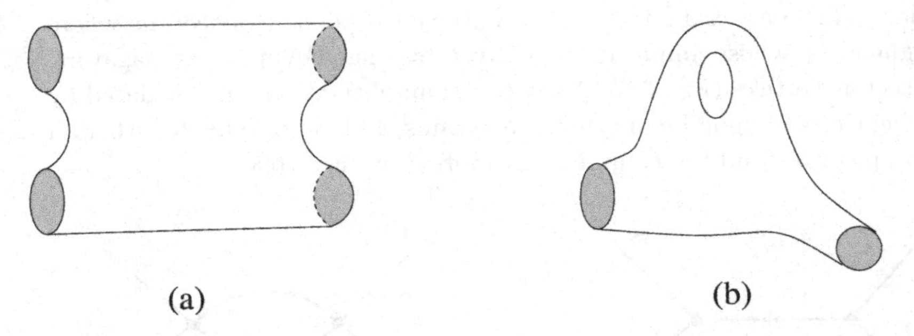

Fig. 3.3 Feynman diagrams in (closed) string theory representing (a) tree-level four-string and (b) one-loop two-string scattering. Loops are associated with handles on the string worldsheet.

To analyse more carefully these Feynman diagrams, it is convenient again to Wick rotate to Euclidean signature so that the worldsheet metric γ_{ab} is positive definite. Then the various topologies that arise in string perturbation theory are *completely* understood. The classification of two-dimensional Euclidean surfaces is a long-solved problem in mathematics. The schematic structure of a generic string scattering amplitude is then given by a path integral of the form

$$\mathcal{A} = \int \mathcal{D}\gamma_{ab}(\xi)\ \mathcal{D}x^\mu(\xi)\ \mathrm{e}^{-S[x,\gamma]}\ \prod_{i=1}^{n_c}\ \int_M \mathrm{d}^2\xi_i\ V_{\alpha_i}(\xi_i)\ \prod_{j=1}^{n_o}\ \int_{\partial M} \mathrm{d}l_j\ V_{\beta_j}(l_j)\ . \tag{3.43}$$

Here γ_{ab} is the metric on the string worldsheet M, $S[x,\gamma]$ is the (ungauged) Polyakov action (2.24), V_{α_i} is the vertex operator that describes the emission or absorption of a closed string state of type α_i from the interior of M, and V_{β_j} is the vertex operator that describes the emission or absorption of an open string state of type β_j from the boundary ∂M of the string worldsheet M.

By using conformal invariance, the amplitudes $\mathcal{A}$ reduce to integrals over conformally inequivalent worldsheets, which at a given topology are described by a finite number of complex parameters called "moduli". In this representation the required momentum integrations are already done. The amplitudes can be thereby recast as *finite* dimensional integrals over the "moduli space of M". The finite dimension $\mathcal{N}$ of this space is given by the number

$$\mathcal{N} = 3(2h + b - 2) + 2n_c + n_o \ , \tag{3.44}$$

where h is the number of "handles" of M, b is its number of boundaries, and n_c and n_o are respectively, as above, the number of closed and open string state insertions on M. As described before, the string coupling g_s is dynamically determined by the worldsheet dilaton field Φ.

3.5 Chan–Paton Factors

We will now describe how to promote the photon fields living at the endpoints of open strings to *non-abelian* gauge field degrees of freedom. For this, we attach non-dynamical "quark" (and "anti-quark") degrees of freedom[1] to the endpoints of an open string in a way which preserves both spacetime Poincaré invariance and worldsheet conformal invariance (Fig. 3.4). In addition to the usual Fock space labels of a string state, we demand that each end of the string be in a state i or j. We further demand that the Hamiltonian of the states $i = 1, \ldots, N$ is 0, so that they stay in the state that we originally put them in for all time. In other words, these states correspond to "background" degrees of freedom. We may then decompose an open string wavefunction $|k; ij\rangle$ in a basis λ^a_{ij} of $N \times N$ matrices as

$$|k; a\rangle = \sum_{i,j=1}^{N} |k; ij\rangle \ \lambda^a_{ij} \ . \tag{3.45}$$

These matrices are called "Chan–Paton factors" [Chan and Paton (1969)]. By the operator-state correspondence, all open string vertex operators also carry such factors.

[1] The "quark" terminology here is only historical, as the strings we are discussing here are now known not to be the long-sought QCD strings thought to be responsible for binding quarks together.

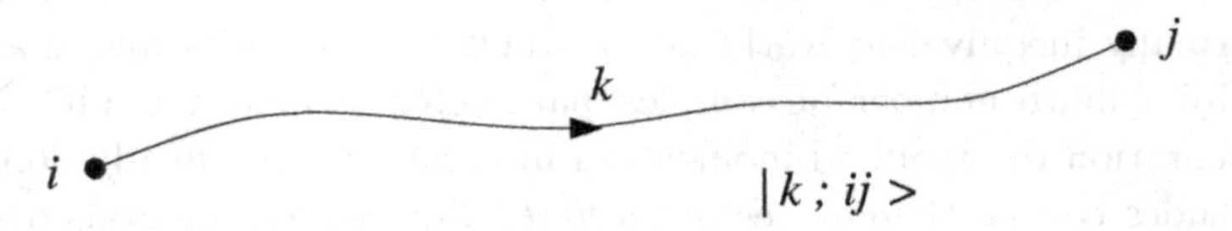

Fig. 3.4 The state $|k;ij\rangle$ obtained by attaching quark labels $i,j = 1,\dots,N$ to the endpoints of an open string. Only the momentum label k of the Fock states is indicated explicitly in the notation.

All open string scattering amplitudes contain traces of products of Chan–Paton factors. As an example, consider the amplitude depicted by the Feynman diagram of Fig. 3.5. Summing over all possible states involved in tying up the ends produces the overall factor

$$\sum_{i,j,k,l} \lambda^1_{ij}\,\lambda^2_{jk}\,\lambda^3_{kl}\,\lambda^4_{li} = \mathrm{Tr}\,\left(\lambda^1\,\lambda^2\,\lambda^3\,\lambda^4\right) \ . \tag{3.46}$$

The amplitude is therefore invariant under a *global* $U(N)$ worldsheet symmetry

$$\lambda^a \ \longmapsto \ U\,\lambda^a\,U^{-1} \ , \quad U \in U(N) \ , \tag{3.47}$$

under which the endpoint i of the string transforms in the fundamental $\mathbf{N}$ representation of the $N\times N$ unitary group $U(N)$, while endpoint j, due to the orientation reversal between the two ends of the string, transforms in the anti-fundamental representation $\overline{\mathbf{N}}$. The corresponding massless vector vertex operator

$$V^{a,\mu}_{ij} = \int \mathrm{d}l\ \lambda^a_{ij}\ {}^{\circ}_{\circ}\,\partial_\parallel x^\mu\ \mathrm{e}^{\,\mathrm{i}\,k\cdot x}\,{}^{\circ}_{\circ} \tag{3.48}$$

thereby transforms under the adjoint $\mathbf{N}\otimes\overline{\mathbf{N}}$ representation of $U(N)$,

$$V^{a,\mu} \ \longmapsto \ U\,V^{a,\mu}\,U^{-1} \ . \tag{3.49}$$

Thus the global $U(N)$ symmetry of the worldsheet field theory is promoted to a local $U(N)$ *gauge* symmetry in spacetime, because we can make a different $U(N)$ rotation at separate points $x^\mu(\tau,\sigma)$ in spacetime. It is in this way that we can promote the photon fields at open string ends to *non-abelian* gauge fields.

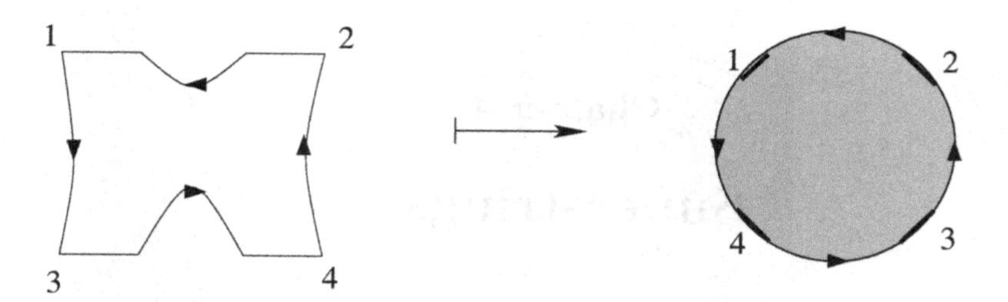

Fig. 3.5 A tree-level four-point open string scattering amplitude with Chan–Paton factors. The right end of string 1 is in the same state as the left end of string 2, and so on, because the Chan–Paton degrees of freedom are non-dynamical. A conformal transformation can be used to map this open string graph onto a disc diagram.

Chapter 4

Superstrings

We will now generalize the bosonic string to include supersymmetry and hence discuss superstrings. After some words of motivation for doing this, we will simply repeat the analysis of the previous two chapters within this new setting, and the constructions will all be the same, but with more rich features arising. After constructing the superstring spectrum of physical states, we will discuss a consistent truncation of it which removes the unphysical degrees of freedom present in the bosonic string spectrum, while at the same time renders the quantum superstring theory supersymmetric in target space. We shall also work out in some detail a one-loop calculation which gives an explicit example of the constructions, and also introduces some important new concepts such as modular invariance.

4.1 Motivation

We will now add fermions to the bosonic string to produce quite naturally supersymmetry, and hence a supersymmetric or spinning string, called "superstring" for short. There are two main reasons why we want to do this:

- Bosonic string theory is sick, because its spectrum of quantum states contains a tachyon, which signals an unstable vacuum.
- Bosonic string theory contains no fermions in its quantum spectrum, and so it has no hope for phenomenological implications. As far as nature is concerned, bosonic strings cannot be the whole story.

There are two ways to introduce supersymmetry in string theory:

- *Ramond–Neveu–Schwarz (RNS) Formalism:* This is the original approach that arose within the context of dual resonance models around 1971 [Neveu and Schwarz (1971); Ramond (1971)]. It uses two-dimensional worldsheet supersymmetry, and it requires the "Gliozzi–Scherck–Olive (GSO) projection" [Gliozzi, Scherk and Olive (1976)] to eventually realize *spacetime* supersymmetry, a spectrum free from tachyons, and also modular invariance.
- *Light-Cone Green–Schwarz Formalism:* In this approach, which emerged around 1981 [Green and Schwarz (1981)], spacetime supersymmetry is explicit from the onset, and it avoids having to use the GSO projection. However, unlike the RNS approach, this superstring theory cannot be easily quantized in a fully covariant way (as its name suggests, it relies heavily on light-cone gauge fixing of the spacetime coordinates). The covariant quantization of the Green–Schwarz superstring has been carried out recently in [Berkovits (2002)].

Here we will only deal with the RNS formalism, which is technically much simpler to deal with and which utilizes many of the techniques of the previous two chapters that we have now become familiar with.

4.2 The RNS Superstring

We start with the gauge-fixed Polyakov action describing d free, massless scalar fields $x^\mu(\tau,\sigma)$. We now add d free, massless Majorana spinors $\psi^\mu(\tau,\sigma)$ on the string worldsheet which transform as d-dimensional vectors under Lorentz transformations in the target spacetime. Again, the worldsheet is the cylinder $-\infty < \tau < \infty$, $0 \le \sigma < 2\pi$ for the closed string, and the strip $-\infty < \tau < \infty$, $0 \le \sigma \le \pi$ for the open string. The worldsheet action in the conformal gauge takes the form [Polyakov (1981b)]

$$\boxed{S = -\frac{T}{2} \int \mathrm{d}^2\xi \left(\partial_a x^\mu \, \partial^a x_\mu - \mathrm{i} \overline{\psi}^\mu \, \rho^a \, \partial_a \psi_\mu \right) ,} \tag{4.1}$$

where the second term in the action is the usual Dirac kinetic term for the fermion fields.

Here ρ^a, $a = 0, 1$ are 2×2 Dirac matrices which in a convenient basis for the worldsheet spinors can be taken to be

$$\rho^0 = \begin{pmatrix} 0 & -\mathrm{i} \\ \mathrm{i} & 0 \end{pmatrix} , \quad \rho^1 = \begin{pmatrix} 0 & \mathrm{i} \\ \mathrm{i} & 0 \end{pmatrix} , \tag{4.2}$$

and they satisfy the worldsheet Dirac algebra

$$\{\rho^a \,,\, \rho^b\} = -2\,\eta^{ab} \, . \tag{4.3}$$

In this basis, the fermion field

$$\psi = \begin{pmatrix} \psi_- \\ \psi_+ \end{pmatrix} \tag{4.4}$$

is a two-component Majorana spinor, $\psi_\pm^* = \psi_\pm$, in order to keep the action (4.1) real. Furthermore, the two-dimensional Dirac term then becomes

$$\overline{\psi} \cdot \rho^a \, \partial_a \psi = \psi_- \cdot \partial_+ \psi_- + \psi_+ \cdot \partial_- \psi_+ \, , \tag{4.5}$$

where the derivatives $\partial_\pm$ are defined in (2.37). The equations of motion for the fermion fields are therefore given by the massless Dirac equation in two dimensions,

$$\partial_+ \psi_-^\mu = \partial_- \psi_+^\mu = 0 \, . \tag{4.6}$$

Thus, the Majorana–Weyl fermions ψ_-^μ describe right-movers while ψ_+^μ describe left-movers. The equations of motion and constraints for the x^μ's are the same as before. The supersymmetry of the field theory defined by the action (4.1) is left as an exercise.

Exercise 4.1. **(a)** *Show that the gauge-fixed fermionic extension (4.1) of the Polyakov action is invariant under the global, infinitesimal worldsheet supersymmetry transformations*

$$\delta_\epsilon x^\mu = \overline{\epsilon}\, \psi^\mu \, ,$$
$$\delta_\epsilon \psi^\mu = -\mathrm{i}\, \rho^a \, \partial_a x^\mu \, \epsilon \, ,$$

with ϵ a constant, anticommuting two-component spinor. Since this transformation acts on both left-moving and right-moving sectors, the worldsheet field theory is said to have "(1, 1) supersymmetry" [Figueroa-O'Farrill (2001)].

(b) *Show that the conserved worldsheet Noether current J_a associated with this symmetry is the fermionic supercurrent*

$$J_a = \frac{1}{2}\, \rho^b \, \rho_a \, \psi^\mu \, \partial_b x_\mu \, .$$

(c) *Show that $\rho^a \, J_a = 0$, so that some components of J_a vanish.*

We can also easily work out the modification of the worldsheet energy–momentum tensor. Using Exercise 4.1 (c), we arrive altogether at the mass-shell constraints

$$\boxed{\begin{aligned} T_{\pm\pm} &\equiv (\partial_\pm x)^2 + \frac{\mathrm{i}}{2}\,\psi_\pm \cdot \partial_\pm \psi_\pm \;=\; 0 \;, \\ J_\pm &\equiv \psi_\pm \cdot \partial_\pm x \;=\; 0 \;. \end{aligned}} \tag{4.7}$$

The second constraint comes from the locally supersymmetric form of the ungauged Polyakov action.

4.2.1 *Mode Expansions*

The mode decompositions for $x^\mu(\tau,\sigma)$ are exactly the same as before. We now need to consider boundary conditions for the free fermionic fields $\psi^\mu(\tau,\sigma)$. Let us first consider the case of open strings. Then the variational principle for the Polyakov action (4.1) requires

$$\psi_+ \cdot \delta\psi_+ - \psi_- \cdot \delta\psi_- = 0 \qquad \text{at} \quad \sigma = 0, \pi \;. \tag{4.8}$$

The Dirac equations of motion thereby admit two possible boundary conditions consistent with Lorentz invariance, namely $\psi_+ = \pm\,\psi_-$ at $\sigma = 0, \pi$ (so that also $\delta\psi_+ = \pm\,\delta\psi_-$ there). The overall relative sign between the fields ψ_- and ψ_+ is a matter of convention, so by redefining ψ_+ if necessary, we may without loss of generality take

$$\psi_+^\mu(\tau,0) = \psi_-^\mu(\tau,0) \;. \tag{4.9}$$

This still leaves two possibilities at the other endpoint $\sigma = \pi$, which are called respectively Ramond (R) and Neveu–Schwarz (NS) boundary conditions:

$$\boxed{\begin{aligned} \psi_+^\mu(\tau,\pi) &= \psi_-^\mu(\tau,\pi) \qquad (\mathrm{R}) \;, \\ \psi_+^\mu(\tau,\pi) &= -\psi_-^\mu(\tau,\pi) \qquad (\mathrm{NS}) \;. \end{aligned}} \tag{4.10}$$

We shall see later on that the R sector will give particles which are spacetime fermions while the NS sector will yield spacetime bosons. The mode decompositions of the Dirac equation (4.6) allow us to express the general solutions as the Fourier series

$$\psi_\pm(\tau,\sigma) = \frac{1}{\sqrt{2}} \sum_r \psi_r^\mu \, \mathrm{e}^{-\mathrm{i}\, r(\tau\pm\sigma)} \, , \\ r = 0, \pm 1, \pm 2, \ldots \quad (\mathrm{R}) \\ = \pm\frac{1}{2}, \pm\frac{3}{2}, \ldots \quad (\mathrm{NS}) \, , \tag{4.11}$$

where the Majorana condition requires

$$\psi_{-r}^\mu = (\psi_r^\mu)^* \; . \tag{4.12}$$

Note that only the R sector gives a zero mode ψ_0^μ.

The closed string sector is analogous, but now we impose periodic or anti-periodic boundary conditions on each component of ψ^μ separately to get the mode expansions

$$\psi_+^\mu(\tau,\sigma) = \sum_r \tilde{\psi}_r^\mu \, \mathrm{e}^{-2\mathrm{i}\, r(\tau+\sigma)} \, , \quad \psi_-^\mu(\tau,\sigma) = \sum_r \psi_r^\mu \, \mathrm{e}^{-2\mathrm{i}\, r(\tau-\sigma)} \, , \tag{4.13}$$

with the mode index r constrained according to (4.11). Corresponding to the different pairings between left-moving and right-moving modes, there are now four distinct closed string sectors that can be grouped together according to the spacetime boson-fermion parity of their quantum states:

Bosons	Fermions
NS – NS	NS – R
R – R	R – NS

(4.14)

Furthermore, the closed string mode expansions of the physical constraints are given by

$$T_{--}(\xi^-) = \sum_{n=-\infty}^{\infty} L_n \, \mathrm{e}^{-2\mathrm{i}\, n\xi^-} \, , \\ J_-(\xi^-) = \sum_r G_r \, \mathrm{e}^{-2\mathrm{i}\, r\xi^-} \, , \tag{4.15}$$

where

$$
\boxed{
\begin{aligned}
L_n &= \frac{1}{2} \sum_{m=-\infty}^{\infty} \alpha_{n-m} \cdot \alpha_m + \frac{1}{4} \sum_r (2r - n)\, \psi_{n-r} \cdot \psi_r \ , \\
G_r &= \sum_{m=-\infty}^{\infty} \alpha_m \cdot \psi_{r-m} \ ,
\end{aligned}
}
\tag{4.16}
$$

and α_m^μ are the bosonic oscillators for the mode expansion of $x^\mu(\tau, \sigma)$. From the $+$ components we similarly get mode operators $\tilde{L}_n$ and $\tilde{G}_r$.

4.3 The Superstring Spectrum

As we did with the bosonic string, we can easily proceed to the canonical quantization of the free two-dimensional field theory (4.1). The fermionic modifications are straightforward to obtain and are again left as an exercise.

Exercise 4.2. **(a)** *Starting from the action (4.1), show that, in addition to the bosonic oscillator commutators given in Exercise 3.1 (b), canonical quantization leads to the anti-commutators*

$$
\{\psi_r^\mu \, , \, \psi_s^\nu\} = \delta_{r+s,0}\, \eta^{\mu\nu} \ .
$$

(b) *Show that the operators L_n and G_r generate the $N = 1$ supersymmetric extension of the Virasoro algebra:*

$$
\begin{aligned}
[L_n, L_m] &= (n - m)\, L_{n+m} + \frac{c}{12} \left(n^3 - n\right) \delta_{n+m,0} \ , \\
[L_n, G_r] &= \frac{1}{2} \left(n - 2r\right) G_{n+r} \ , \\
\{G_r, G_s\} &= 2\, L_{r+s} + \frac{c}{12} \left(4r^2 - 1\right) \delta_{r+s,0} \ ,
\end{aligned}
$$

where $c = d + \frac{d}{2}$ is the total contribution to the conformal anomaly.

(c) *Show that the conserved angular momentum is given by $J^{\mu\nu} = J_\alpha^{\mu\nu} + K^{\mu\nu}$, where $J_\alpha^{\mu\nu}$ is the contribution from the bosonic modes (Exercise 3.2) and*

$$
K^{\mu\nu} = -\mathrm{i} \sum_{r \geq 0} \left(\psi_{-r}^\mu\, \psi_r^\nu - \psi_{-r}^\nu\, \psi_r^\mu\right) \ .
$$

The canonical anticommutators obtained in Exercise 4.2 (a) are just the standard ones in the canonical quantization of free Fermi fields $\psi^\mu(\tau,\sigma)$. The basic structure $\{\psi_r^{\mu\,\dagger},\psi_r^\mu\} = 1$ is *very* simple, as it describes a two-state system spanned by vectors $|0\rangle$ and $|1\rangle$ with $\psi_r^\mu|0\rangle = 0$, $\psi_r^{\mu\,\dagger}|0\rangle = |1\rangle$ for $r > 0$. The ψ_r^μ with $r < 0$ may then be regarded as raising operators, and as lowering operators for $r > 0$. The full state space is the free tensor product of the bosonic and fermionic Hilbert spaces constructed in this way.

By using precisely the same techniques as for the bosonic string in the previous chapter, we can again fix the spacetime dimension d and the normal ordering constant a. The "superstring critical dimension" is now

$$\boxed{d = 10\ .} \tag{4.17}$$

The bosonic oscillators α_n^μ contribute, as before, a regulated Casimir energy $-\frac{d-2}{24}$, while the fermionic modes ψ_r^μ yield minus that value for integer r and minus half that value $\frac{d-2}{48}$ for half-integer r. The normal ordering constant is thereby found to be

$$\boxed{a = \begin{cases} 0 & \text{(R)} \\ \frac{1}{2} & \text{(NS)}\ . \end{cases}} \tag{4.18}$$

The physical state conditions are given by the super-Virasoro constraints

$$\begin{aligned} (L_0 - a)|\text{phys}\rangle &= 0\ , \\ L_n|\text{phys}\rangle &= 0\ , \quad n > 0\ , \\ G_r|\text{phys}\rangle &= 0\ , \quad r > 0\ . \end{aligned} \tag{4.19}$$

The $L_0 = a$ constraint yields the open string mass formula

$$\boxed{m^2 = \frac{1}{\alpha'}\Big(N - a\Big)\ ,} \tag{4.20}$$

where the total level number is given by

$$N = \sum_{n=1}^{\infty} \alpha_{-n}\cdot\alpha_n + \sum_{r>0} r\,\psi_{-r}\cdot\psi_r\ . \tag{4.21}$$

Aside from the change in values of a in (4.18) and the definition (4.21), this is the same mass formula as obtained in the previous chapter.

4.3.1 *The Open Superstring Spectrum*

The open string spectrum of states has the two independent NS and R sectors, which we will study individually.

NS Sector : The NS ground state $|k;0\rangle_{\rm NS}$ satisfies

$$\begin{aligned} \alpha_n^\mu|k;0\rangle_{\rm NS} = \psi_r^\mu|k;0\rangle_{\rm NS} &= 0 \ , \quad n,r>0 \ , \\ \alpha_0^\mu|k;0\rangle_{\rm NS} &= \sqrt{2\alpha'}\,k^\mu|k;0\rangle_{\rm NS} \ . \end{aligned} \tag{4.22}$$

This sector of the Hilbert space of physical states of the RNS superstring is then a straightforward generalization of the bosonic string spectrum. In particular, the vacuum $|k;0\rangle_{\rm NS}$ has $m^2=-\frac{1}{2\alpha'}$ and is tachyonic. We'll soon see how to eliminate it from the physical spectrum.

The first excited levels in this sector contain the massless states given by $\psi^\mu_{-\frac{1}{2}}|k;0\rangle_{\rm NS}$, $m^2=0$. They are vectors of the transverse $SO(8)$ rotation group which corresponds to the Little group of $SO(1,9)$ that leaves the light-cone momentum invariant. They describe the eight physical polarizations of the massless, open string photon field $A_\mu(x)$ in ten spacetime dimensions. All states in the NS sector are spacetime bosons, because they transform in appropriate irreducible representations of $SO(8)$ in the decompositions of tensor products of the vectorial one.

R Sector : In the Ramond sector there are zero modes ψ_0^μ which satisfy the ten dimensional Dirac algebra

$$\{\psi_0^\mu\,,\,\psi_0^\nu\}=\eta^{\mu\nu} \ . \tag{4.23}$$

Thus the ψ_0's should be regarded as Dirac matrices, $\psi_0^\mu=\frac{1}{\sqrt{2}}\Gamma^\mu$, and in particular they are finite dimensional operators. All states in the R sector should be ten dimensional spinors in order to furnish representation spaces on which these operators can act. We conclude that all states in the R sector are spacetime fermions.

The zero mode part of the fermionic constraint $J_-=0$ in (4.7) gives a wave equation for fermionic strings in the R sector known as the "Dirac–Ramond equation"

$$G_0|{\rm phys}\rangle_{\rm R}=0 \ , \tag{4.24}$$

where

$$G_0 = \alpha_0 \cdot \psi_0 + \sum_{n \neq 0} \alpha_{-n} \cdot \psi_n \tag{4.25}$$

obeys (Exercise 4.2 (b))

$$G_0^2 = L_0 \ . \tag{4.26}$$

Note that the zero mode piece $\alpha_0 \cdot \psi_0$ in (4.25) is precisely the spacetime Dirac operator $\partial\!\!\!/ = \Gamma^\mu \partial_\mu$ in momentum space, because $\alpha_0^\mu \propto p_0^\mu$ and $\psi_0^\mu \propto \Gamma^\mu$. Thus the R sector fermionic ground state $|\psi^{(0)}\rangle_{\rm R}$, defined by

$$\alpha_n^\mu |\psi^{(0)}\rangle_{\rm R} = \psi_n^\mu |\psi^{(0)}\rangle_{\rm R} = 0 \ , \quad n > 0 \ , \tag{4.27}$$

satisfies the massless Dirac wave equation

$$\alpha_0 \cdot \psi_0 |\psi^{(0)}\rangle_{\rm R} = 0 \ . \tag{4.28}$$

It follows that the fermionic ground state of the superstring is a massless Dirac spinor in ten dimensions. However, at present, it has too many components to form a supersymmetric multiplet with the bosonic ground state of the NS sector. In ten spacetime dimensions, it is known that one can impose both the Majorana and Weyl conditions simultaneously on spinors, so that the ground state $|\psi^{(0)}\rangle_{\rm R}$ can be chosen to have a definite spacetime chirality. We will denote the two possible chiral ground states into which it decomposes by $|a\rangle_{\rm R}$ and $|\overline{a}\rangle_{\rm R}$, where $a, \overline{a} = 1, \ldots, 8$ are spinor indices labelling the two inequivalent, irreducible Majorana–Weyl spinor representations of $SO(8)$.

4.3.2 *The Closed Superstring Spectrum*

The spectrum of closed strings is obtained by taking tensor products of left-movers and right-movers, each of which is very similar to the open superstring spectrum obtained above, and by again using appropriate level-matching conditions. Then, as mentioned before, there are four distinct sectors (see (4.14)). In the NS–NS sector, the lowest lying level is again a closed string tachyon. The totality of *massless* states, which are picked up in the field theory limit $\alpha' \to 0$, can be neatly summarized according to their transformation properties under $SO(8)$ and are given in Table 4.1. In this way the closed superstring sector endows us now with spacetime "supergravity fields".

Table 4.1 The massless states of the closed superstring. The fields in the last column are identified according to the irreducible $SO(8)$ representations of the third column. The subscript v denotes vector and the subscript s spinor representation. Ψ_μ and Ψ'_μ are spin $\frac{3}{2}$ gravitino fields, while λ and λ' are spin $\frac{1}{2}$ dilatino fields. These fermionic states all have the same helicity. The Ramond–Ramond sector will be described in more detail in the next chapter.

SECTOR	BOSON/FERMION?	$SO(8)$ REPRESENTATION	MASSLESS FIELDS
NS–NS	boson	$\mathbf{8}_v \otimes \mathbf{8}_v = \mathbf{35} \oplus \mathbf{28} \oplus \mathbf{1}$	$g_{\mu\nu}$, $B_{\mu\nu}$, Φ
NS–R	fermion	$\mathbf{8}_v \otimes \mathbf{8}_s = \mathbf{8}_s \oplus \mathbf{56}_s$	Ψ_μ , λ
R–NS	fermion	$\mathbf{8}_s \otimes \mathbf{8}_v = \mathbf{8}_s \oplus \mathbf{56}_s$	Ψ'_μ , λ'
R–R	boson	$\mathbf{8}_s \otimes \mathbf{8}_s = p -$ forms	Ramond–Ramond fields

4.4 The GSO Projection

The superstring spectrum admits a consistent truncation, called the "GSO projection", which is necessary for consistency of the interacting theory. For example, it will remedy the situation with the unwanted NS sector tachyon which has $m^2 = -\frac{1}{2\alpha'}$. Again we will examine this operation separately in the NS and R sectors of the worldsheet theory.

NS Sector : The GSO projection $P_{\rm GSO}$ here is defined by keeping states with an odd number of ψ oscillator excitations, and removing those with an even number. This is tantamount to replacing physical states according to

$$|{\rm phys}\rangle_{\rm NS} \longmapsto P_{\rm GSO}|{\rm phys}\rangle_{\rm NS} \tag{4.29}$$

with

$$P_{\rm GSO} = \frac{1}{2}\left(1-(-1)^F\right) , \tag{4.30}$$

where

$$F = \sum_{r>0} \psi_{-r}\cdot\psi_r \tag{4.31}$$

is the "fermion number operator" which obeys

$$\left\{(-1)^F\, ,\, \psi^\mu\right\} = 0 \ . \tag{4.32}$$

Thus only half-integer values of the level number (4.21) are possible, and so the spectrum of allowed physical masses are integral multiples of $\frac{1}{\alpha'}$, $m^2 = 0, \frac{1}{\alpha'}, \frac{2}{\alpha'}, \ldots$. In particular, the bosonic ground state is now massless, and the spectrum no longer contains a tachyon (which has fermion number $F = 0$).

R Sector : Here we use the same formula (4.30), but now we define the Klein operator $(-1)^F$ by

$$(-1)^F = \pm\Gamma^{11}\cdot(-1)^{\sum_{r\geq 1}\psi_{-r}\cdot\psi_r}\ , \tag{4.33}$$

where

$$\Gamma^{11} = \Gamma^0\,\Gamma^1\cdots\Gamma^9 \tag{4.34}$$

is the ten dimensional chirality operator with

$$\left(\Gamma^{11}\right)^2 = 1\ ,\quad \{\Gamma^\mu\,,\,\Gamma^{11}\} = 0\ . \tag{4.35}$$

The operators $1 \pm \Gamma^{11}$ project onto spinors of opposite spacetime chirality. As we will see below, this chirality projection will guarantee spacetime supersymmetry of the physical superstring spectrum. Although here the choice of sign in (4.33), corresponding to different chirality projections on the spinors, is merely a matter of convention, we will see in the next chapter that it significantly affects the physical properties of the *closed* string sector.

4.4.1 *Spacetime Supersymmetry*

Let us now examine the massless spectrum of the GSO-projected superstring theory. The ground state (NS) boson is described by the state $\zeta\cdot\psi_{-\frac{1}{2}}|k;0\rangle_{\rm NS}$. As before, it is a massless vector and has $d-2=8$ physical polarizations. The ground state (R) fermion is described by the state $P_{\rm GSO}|\psi^{(0)}\rangle_{\rm R}$. It is a massless Majorana–Weyl spinor which has $\frac{1}{4}\cdot 2^{d/2}=8$ physical polarizations. Thus the ground states have an equal number of bosons and fermions, as required for supersymmetry. In particular, they form the $\mathbf{8}_v\oplus\mathbf{8}_s$ vector supermultiplet of the $d=10$, $N=1$ supersymmetry algebra. In fact, this state yields the pair of fields that forms the vector supermultiplet of supersymmetric Yang–Mills theory in ten spacetime dimensions, the extension to $U(N)$ gauge symmetry with Chan–Paton factors being straightforward. It can also be shown that the entire interacting superstring theory (beyond the massless states) now has spacetime supersymmetry [Green and Schwarz (1982)]. Thus in addition to removing the tachyonic instabilities of the vacuum, the GSO projection has naturally provided us with a target space supersymmetric theory as an additional bonus. We will see another remarkable consequence of this projection in the next section.

4.5 Example: One-Loop Vacuum Amplitude

Example 4.1. We will now go through an explicit calculation of a loop diagram in superstring theory, as an illustration of some of the general features of quantum superstring theory that we have developed thus far. This example will also introduce a very important property of superstring perturbation theory that is absolutely essential to the overall consistency of the model. We will need some identities involving traces over the infinite families of oscillators that are present in the quantum string mode expansions, which are left as an exercise that can be straightforwardly done by carefully tabulating states of the oscillator algebras.

Exercise 4.3. **(a)** *Given a set of one-dimensional bosonic oscillators* α_n*, show that for any complex number* $q \neq 1$,

$$\mathrm{tr}\left(q^{\sum_{n=1}^{\infty}\alpha_{-n}\,\alpha_n}\right) = \prod_{n=1}^{\infty}\frac{1}{1-q^n}\ .$$

(b) *Given a set of one-dimensional fermionic oscillators* ψ_n*, show that*

$$\mathrm{tr}\left(q^{\sum_{n=1}^{\infty}\psi_{-n}\,\psi_n}\right) = \prod_{n=1}^{\infty}\left(1+q^n\right)\ .$$

We will now compute the one-loop closed string vacuum diagram. At one-loop order, the Feynman graph describing a closed string state which propagates in time ξ^0 and returns back to its initial state is a donut-shaped surface, i.e. a two-dimensional torus (Fig. 4.1). The amplitude is given by the Polyakov path integral (3.43), which, by conformal invariance, sums over only conformally inequivalent tori. Let us first try to understand this point geometrically. We can specify the torus by giving a flat metric, which we will take to be $\eta_{ab} = \delta_{ab}$ in fixing the conformal gauge, along with a complex structure $\tau \in \mathbb{C}$, with $\mathrm{Im}(\tau) > 0$. The complex number τ specifies the shape of the torus, which cannot be changed by a conformal transformation of the metric (nor any local change of coordinates). This is best illustrated by the "parallelogram representation" of the torus in Fig. 4.2.

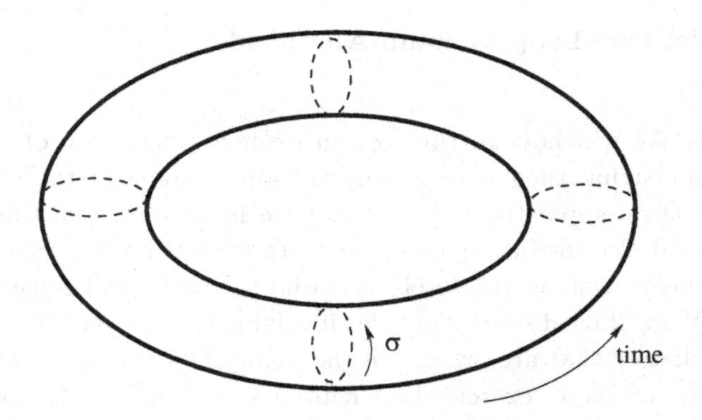

Fig. 4.1 The one-loop closed string vacuum diagram is a two dimensional torus.

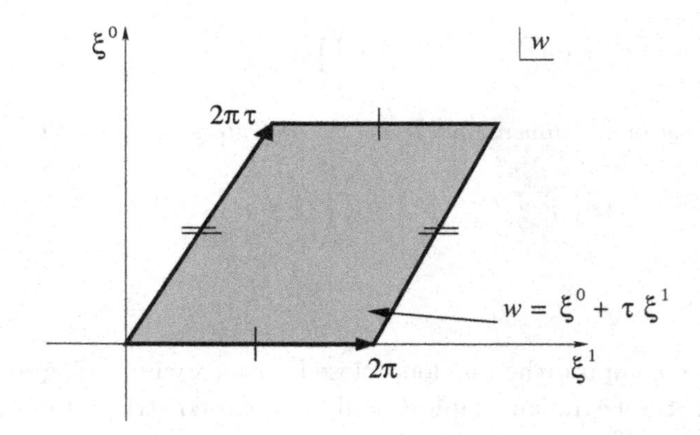

Fig. 4.2 The parallelogram representation of the torus. Opposite edges of the parallelogram are periodically identified, i.e. $0 \le \xi^0 < 2\pi$, $0 \le \xi^1 < 2\pi$. Equivalently, it can be viewed as the region of the complex w-plane obtained from the identifications $w \sim w + 2\pi n$ and $w \sim w + 2\pi m\tau$ for all integers n and m.

Lots of different τ's define the same torus. For example, from the equivalence relation description of Fig. 4.2 it is clear that τ and $\tau + 1$ give the same torus. The full family of equivalent tori can be reached from any τ by "modular transformations" which are combinations of the operations

$$T : \tau \longmapsto \tau + 1 \ , \quad S : \tau \longmapsto \ -\frac{1}{\tau} \ . \tag{4.36}$$

The transformations S and T obey the relations

$$S^2 = (ST)^3 = 1 \tag{4.37}$$

and they generate the "modular group" $SL(2,\mathbb{Z})$ of the torus acting on τ by the discrete linear fractional transformations

$$\tau \longmapsto \frac{a\tau + b}{c\tau + d} \ , \quad \begin{pmatrix} a & b \\ c & d \end{pmatrix} \in SL(2,\mathbb{Z}) \ , \tag{4.38}$$

where the $SL(2,\mathbb{Z})$ condition restricts $a,b,c,d \in \mathbb{Z}$ with $ad - bc = 1$. We conclude that the "sum" over τ should be restricted to a "fundamental modular domain" $\mathcal{F}$ in the upper complex half-plane. Any point outside $\mathcal{F}$ can be mapped into it by a modular transformation. A convenient choice of fundamental region is depicted in Fig. 4.3.

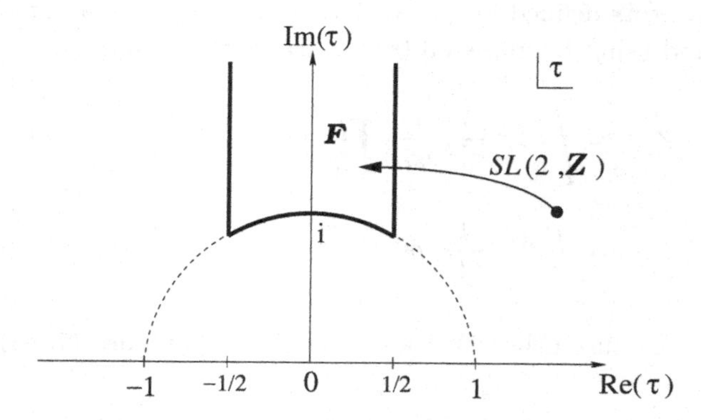

Fig. 4.3 A fundamental modular domain $\mathcal{F}$ for the torus. The interior of this region is invariant under the basis transformations (4.36). Any point outside of $\mathcal{F}$ can be mapped into it by an $SL(2,\mathbb{Z})$ transformation (4.38).

We are now ready to evaluate explicitly the vacuum amplitude. We will begin with the bosonic string. For this, let us consider string propagation on the torus as depicted in Fig. 4.2. A fixed point on the string, which we assume is lying horizontally in the complex w-plane, propagates upwards for a time $\xi^0 = 2\pi\,\mathrm{Im}(\tau) = 2\pi\tau_2$, at which point it also shifts to the right by an amount $\xi^1 = 2\pi\,\mathrm{Re}(\tau) = 2\pi\tau_1$. The time translation is affected by the worldsheet Hamiltonian $H = L_0 + \tilde{L}_0 - 2$, while the shift along the string is affected by the worldsheet momentum $P = L_0 - \tilde{L}_0$. So the bosonic vacuum path integral is given by

$$Z_{\rm bos} \equiv \mathrm{Tr}\left(\mathrm{e}^{-2\pi\tau_2 H}\ \mathrm{e}^{2\pi\,\mathrm{i}\,\tau_1 P}\right) = \mathrm{Tr}\left(q^{L_0-1}\,\overline{q}^{\tilde{L}_0-1}\right) , \tag{4.39}$$

where

$$q \equiv \mathrm{e}^{2\pi\,\mathrm{i}\,\tau} \tag{4.40}$$

and Tr denotes the trace taken by summing over all discrete oscillator states, and by integrating over the continuous zero-mode momenta k^μ and inequivalent τ, along with the appropriate physical state restrictions. Substituting in the mode expansions (2.42), (2.43) for L_0 and $\tilde{L}_0$ gives

$$Z_{\rm bos} = \int\limits_{\mathcal{F}} \mathrm{d}^2\tau\ \frac{1}{q\overline{q}} \int \mathrm{d}^{24}k\ \mathrm{e}^{-\pi\tau_2 k^2/2}\ \mathrm{tr}\left(q^N\,\overline{q}^{\tilde{N}}\right) , \tag{4.41}$$

where tr denotes the trace over just the oscillator states, and N and $\tilde{N}$ are the number operators defined by (3.13). Evaluating the Gaussian momentum integrals and using Exercise 4.3 (a) we arrive finally at

$$\begin{aligned} Z_{\rm bos} &= \int\limits_{\mathcal{F}} \mathrm{d}^2\tau\ \frac{1}{\tau_2^{12}}\,\frac{1}{q\overline{q}} \left|\prod_{n=1}^{\infty}(1-q^n)^{-24}\right|^2 \\ &= \int\limits_{\mathcal{F}} \mathrm{d}^2\tau\ \frac{1}{\tau_2^{12}} \left|\eta(\tau)\right|^{-48} , \end{aligned} \tag{4.42}$$

where we have introduced the "Dedekind function" [Mumford (1983)]

$$\eta(\tau) = q^{1/24} \prod_{n=1}^{\infty}(1-q^n) \tag{4.43}$$

with the modular transformation properties

$$\eta(\tau+1) = \mathrm{e}^{\pi\,\mathrm{i}/12}\,\eta(\tau)\ ,\quad \eta\left(-\frac{1}{\tau}\right) = \sqrt{-\mathrm{i}\,\tau}\ \eta(\tau)\ . \tag{4.44}$$

Exercise 4.4. *Show that the integrand of the partition function (4.42) is "modular invariant", i.e. it is unaffected by $SL(2,\mathbb{Z})$ transformations of τ.*

The modular invariance property of Exercise 4.4 is crucial for the overall consistency of the quantum string theory, because it means that we are correctly integrating over *all* inequivalent tori, and also that we are counting each such torus only once. However, the modular integral (4.42) over $\mathcal{F}$ actually *diverges*, which is a consequence of the tachyonic instability of the bosonic string theory. We shall now examine how this is remedied by the RNS superstring. For this, we will first interpret the Ramond and Neveu–Schwarz sectors of the theory geometrically.

4.5.1 *Fermionic Spin Structures*

Fermions on the torus are specified by a choice of "spin structure", which is simply a choice of periodic or anti-periodic boundary conditions as $\xi^0 \mapsto \xi^0 + 2\pi$ and $\xi^1 \mapsto \xi^1 + 2\pi$ (see Fig. 4.2). There are four possible spin structures in all, which we will denote symbolically by

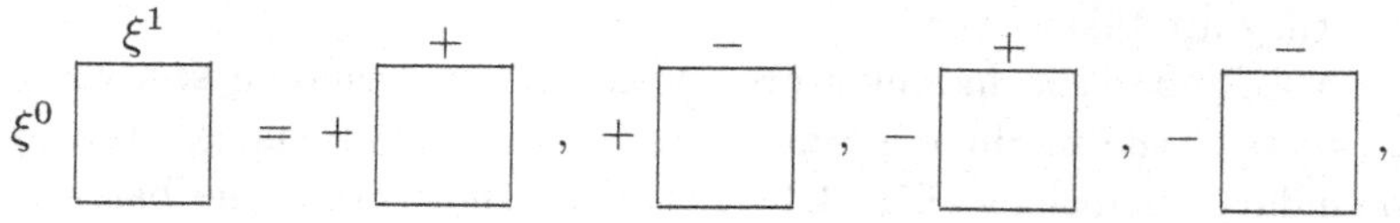

$$\xi^0\overset{\xi^1}{\square} = +\overset{+}{\square}\ , \ +\overset{-}{\square}\ , \ -\overset{+}{\square}\ , \ -\overset{-}{\square}\ , \qquad (4.45)$$

where the squares denote the result of performing the functional integral over fermions with the given fixed spin structure, specified by the labelled sign change $\pm$ of the fermion fields around the given ξ^a cycle of the torus. For the moment we focus our attention only on the right-moving sector. Looking back at the mode expansions (4.11), we see that the NS (resp. R) sector corresponds to anti-periodic (resp. periodic) boundary conditions along the string in the ξ^1 direction. In order to implement periodicity in the time direction ξ^0, we insert the Klein operator $(-1)^F$ in the traces defining the partition function. Because of the anticommutation property (4.32), its insertion in the trace flips the boundary condition around ξ^0, and without it the boundary condition is anti-periodic, in accordance with the standard path integral formulation of finite temperature quantum field theory.[1] The four fermionic spin structure contributions to the path integral are thereby given as

[1] In the path integral approach to quantum statistical mechanics, the Euclidean "time" direction is taken to lie along a circle whose circumference is proportional to the inverse temperature of the system. It is known that free fermions must then be anti-periodic around the time direction in order to reproduce the correct Fermi–Dirac distribution.

$$- \, \square^{+} \equiv \mathrm{Tr}_{\mathrm{R}}\Big(\mathrm{e}^{-2\pi\tau_2 H}\Big)\,, \tag{4.46}$$

$$+ \, \square^{+} = \mathrm{Tr}_{\mathrm{R}}\Big((-1)^F \; \mathrm{e}^{-2\pi\tau_2 H}\Big)\,, \tag{4.47}$$

$$- \, \square^{-} \equiv \mathrm{Tr}_{\mathrm{NS}}\Big(\mathrm{e}^{-2\pi\tau_2 H}\Big)\,, \tag{4.48}$$

$$+ \, \square^{-} = \mathrm{Tr}_{\mathrm{NS}}\Big((-1)^F \; \mathrm{e}^{-2\pi\tau_2 H}\Big)\,, \tag{4.49}$$

where the subscripts on the traces indicate in which sector of the fermionic Hilbert space they are taken over.

Let us now consider the modular properties of the fermionic sector. The action (4.38) of the modular group $SL(2,\mathbb{Z})$ in general changes the fermionic boundary conditions (Fig. 4.4). It can be shown that the basis modular transformations (4.36) mix the various spin structures (4.45) according to

$$\begin{aligned}
T \, : \quad & + \, \square^{+} \longmapsto + \, \square^{+} \,, \qquad - \, \square^{+} \longmapsto - \, \square^{+} \,, \\
& + \, \square^{-} \longmapsto - \, \square^{-} \,, \qquad - \, \square^{-} \longmapsto + \, \square^{-} \,, \\
S \, : \quad & + \, \square^{+} \longmapsto + \, \square^{+} \,, \qquad - \, \square^{+} \longmapsto + \, \square^{-} \,, \\
& + \, \square^{-} \longmapsto - \, \square^{+} \,, \qquad - \, \square^{-} \longmapsto - \, \square^{-} \,.
\end{aligned} \tag{4.50}$$

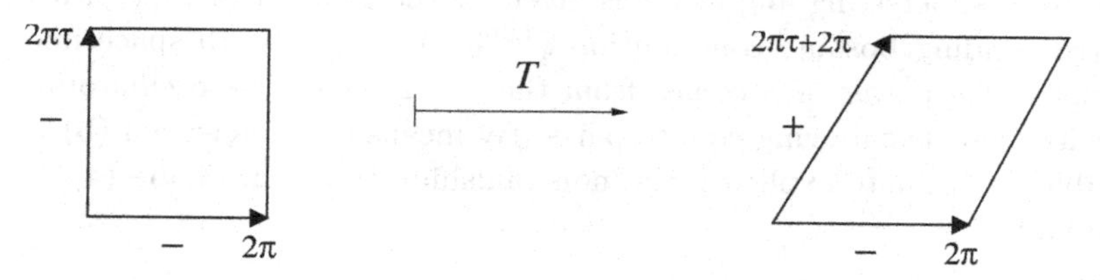

Fig. 4.4 The modular transformation $T : \tau \mapsto \tau + 1$ induces an additional periodic shift from ξ^1 along the worldsheet time direction. It therefore flips the ξ^0 boundary condition whenever the ξ^1 boundary condition is anti-periodic.

Note that the $(+,+)$ spin structure is modular invariant. In fact, the corresponding amplitude (4.47) vanishes, because it contains a Grassmann integral over the constant fermionic zero-modes but the Hamiltonian L_0 is independent of these modes (see (4.16)). This is the only amplitude which contains such fermionic zero-modes. For the remaining spin structures, it follows from (4.50) that their unique modular invariant combination is given up to an overall constant by

$$\overset{-}{-\,\square}\;-\;\overset{-}{+\,\square}\;-\;\overset{+}{-\,\square}\;. \qquad (4.51)$$

The one-loop, modular invariant partition function for the right-moving fermionic contributions is therefore given by this sum, which from (4.46)–(4.49) reads explicitly

$$Z_{\rm RNS} = \frac{1}{2}\,{\rm Tr}_{\rm NS}\left[\left(1-(-1)^F\right)q^{L_0-\frac{1}{2}}\right] - \frac{1}{2}\,{\rm Tr}_{\rm R}\left(q^{L_0}\right)\,. \qquad (4.52)$$

In the first term we recognize the GSO projection operator (4.30), and so using the vanishing of (4.47) we may write the amplitude (4.52) succinctly as a trace over the full right-moving fermionic Hilbert space as

$$Z_{\rm RNS} = {\rm Tr}_{{\rm NS}\oplus{\rm R}}\left(P_{\rm GSO}\,q^{L_0-a}\right)\,. \qquad (4.53)$$

We have thereby arrived at a beautiful interpretation of the GSO projection, which ensures vacuum stability and spacetime supersymmetry of the quantum string theory. Geometrically, it is simply the *modular invariant* sum over spin structures. This interpretation also generalizes to higher-loop amplitudes.

The total superstring amplitude is given by the product of $Z_{\rm RNS}$ with the corresponding bosonic contribution $q^{1/3}\,\eta(\tau)^{-8}$ in $d = 10$ spacetime dimensions (the power of 8 comes from the $d-2$ transverse oscillators), along with their left-moving counterparts. By means of Exercise 4.3 (b) it is possible to evaluate explicitly the non-vanishing traces in (4.46)–(4.49) and we find

$$
\begin{aligned}
-\overset{+}{\Box} &= 16\,q^{2/3}\prod_{n=1}^{\infty}\left(1+q^{2n}\right)^{8} \ , \\
+\overset{-}{\Box} &= q^{-1/3}\prod_{n=1}^{\infty}\left(1-q^{2n-1}\right)^{8} \ , \\
-\overset{-}{\Box} &= q^{-1/3}\prod_{n=1}^{\infty}\left(1+q^{2n-1}\right)^{8} \ .
\end{aligned}
\tag{4.54}
$$

The remarkable feature of the modular invariant combination (4.51) is that it *vanishes* identically, as a consequence of the “Jacobi abstruse identity” [Mumford (1983)]

$$
q^{-1/2}\left[\prod_{n=1}^{\infty}\left(1+q^{n-1/2}\right)^{8}-\prod_{n=1}^{\infty}\left(1-q^{n-1/2}\right)^{8}\right] = 16\prod_{n=1}^{\infty}\left(1+q^{n}\right)^{8} \ .
\tag{4.55}
$$

This remarkable, non-trivial mathematical identity implies that the full superstring vacuum amplitude vanishes, which provides *very* strong evidence in favour of spacetime supersymmetry in string theory in ten dimensions. It simply reflects the fact that the spacetime NS sector bosons and R sector fermions contribute the same way in equal numbers (but with opposite signs due to statistics).[2] This is of course by no means a complete proof of spacetime supersymmetry to all orders of string perturbation theory. For that, one needs to use the Green–Schwarz formalism [Green and Schwarz (1982)], which will not be dealt with here.

[2]When Jacobi discovered his formula (4.55) in 1829, not realizing any of its immediate implications he refered to it as “a very obscure formula”. Some 150 years later string theory provided a natural explanation of it, namely that it is the requirement of spacetime supersymmetry at one-loop order in string perturbation theory.

Chapter 5

Ramond–Ramond Charges and T-Duality

We will now start moving our attention away from perturbative string theory and begin working our way towards describing D-branes. We shall begin by finishing off our analysis of the perturbative superstring spectrum by describing in some detail the R–R sector, which was mostly ignored in the previous chapter. This will lead us to define the notion of Ramond–Ramond charge which will immediately imply the existence of D-branes in superstring theory. To construct them explicitly, we shall describe in detail the phenomenon of T-duality, which is a purely stringy effect whose implications on the spacetime structure of string theory is interesting in its own right. This duality symmetry, like S-duality in gauge theories, is an indispensible tool in string theory. It also yields an example of a stringy property which does not have a counterpart in quantum field theory. We will describe individually the T-duality symmetries of closed strings, open strings, and of superstrings.

5.1 Ramond–Ramond Charges

In this chapter we will for the most part be concerned with introducing some deep reasons and motivation for the appearence of D-branes in superstring theory. Our first step will be to return to the R–R sector of the closed RNS superstring spectrum of the previous chapter, which we will now describe in detail. Recall that for *open* strings, there are two possible Ramond sector GSO projections given by the operators

$$P^{\pm}_{\text{GSO}} = \frac{1}{2}\left(1 \pm \Gamma^{11} \cdot (-1)^{\sum_{r\geq 1} \psi_{-r}\cdot\psi_r}\right) . \tag{5.1}$$

The sign choice in (5.1) selects a Ramond ground state spinor $|\psi\rangle_{\rm R}$ (as well as massive fermions in the spectrum) of a given $\pm$ spacetime chirality:

$$\Gamma^{11}|\psi\rangle_{\rm R} = \pm\,|\psi\rangle_{\rm R}\ . \tag{5.2}$$

The particular chirality is simply a matter of taste, as either choice leads to physically equivalent results. Thus either $P^+_{\rm GSO}$ or $P^-_{\rm GSO}$ is a good GSO projection in the open string sector.

However, the distinction between $P^\pm_{\rm GSO}$ is meaningful in the *closed* string sector, when left-movers and right-movers are combined together with particular choices of chiralities. The GSO projection is performed separately in both sectors, in each of which there is a two-fold ambiguity. Altogether there are four possible choices. Of these, we can identify two of them, by flipping the sign convention of the total mod 2 fermion number operator $(-1)^{F_{\rm L}+F_{\rm R}}$ if necessary. Then, depending on the choice of relative sign in defining the Klein operators $(-1)^{F_{\rm L}}$ and $(-1)^{F_{\rm R}}$, there are two inequivalent possibilities corresponding to the relative chirality between the surviving R-sector Majorana–Weyl spinors $|\psi_{\rm l}\rangle_{\rm R}$ and $|\psi_{\rm r}\rangle_{\rm R}$. Since we have to pick two copies in order to make a closed string, there are two possible string theories that one can construct in this way, called "Type IIA" and "Type IIB":

Type IIA : In this case we take the *opposite* GSO projection on both sides, so that the spinors are of opposite chirality and hence admit the expansions

$$|\psi_{\rm l}\rangle_{\rm R} = \sum_a (\psi_{\rm l})_a\,|a\rangle_{\rm R}\ , \quad |\psi_{\rm r}\rangle_{\rm R} = \sum_{\overline{a}} (\psi_{\rm r})_{\overline{a}}\,|\overline{a}\rangle_{\rm R}\ . \tag{5.3}$$

The resulting theory is non-chiral.

Type IIB : Here we impose the *same* GSO projection on both sides, so that the spinors have the same chirality with

$$|\psi_{\rm l,r}\rangle_{\rm R} = \sum_a (\psi_{\rm l,r})_a\,|a\rangle_{\rm R}\ . \tag{5.4}$$

This leads to a chiral theory.

5.1.1 *Remarks on Superstring Types*

For completeness, we can now give brief definitions and descriptions of the five consistent, perturbative superstring theories in ten spacetime dimensions that were mentioned in Chapter 1.

(1) Type II Superstrings : These are the superstrings that we have been studying thus far. They involve strings with oriented worldsheets (like the torus), and they possess local $N = 2$ spacetime supersymmetry after implementation of the GSO projection (this is the reason for the terminology Type II). While the ground state chiral structure of the Type IIA and IIB theories differ, their massive states are all the same. We will continue to study only Type II superstrings in the remainder of these notes.

(2) Type I Superstrings : These can be obtained from a projection of the Type IIB theory that keeps only the diagonal sum of the two gravitinos Ψ_μ and Ψ'_μ (see Table 4.1, p. 48). This theory has only $N = 1$ spacetime supersymmetry and is a theory of unoriented string worldsheets (like the Möbius strip or the Klein bottle). In the open string sector, quantum consistency (anomaly cancellation) singles out $SO(32)$ as the only possible Chan–Paton gauge group [Green and Schwarz (1984)].

(3) Heterotic Strings : This comprises a heterosis of two string theories [Gross *et al* (1985)], whereby we use the $d = 26$ bosonic string for the left-movers and the $d = 10$ superstring for the right-movers. The remaining 16 right-moving degrees of freedom required by $N = 1$ supersymmetry are "internal" ones which come from a 16-dimensional, even self-dual lattice (these latter two constraints on the lattice are required by modular invariance). There are only two such lattices, corresponding to the weight lattices of the Lie groups $E_8 \times E_8$ and $SO(32)$ (or more precisely its spin cover $Spin(32)/\mathbb{Z}_2$).[1]

5.1.2 *Type II Ramond–Ramond States*

We will now study in some detail the structure of the ground states in both the Type IIA and IIB R–R sectors. Crucial to this analysis will be a number of Γ-matrix identities in ten dimensions, which are given in the following exercise.

[1] $N = 1$ supersymmetric Yang–Mills theory and supergravity in ten spacetime dimensions possess "hexagon" gauge and gravitational anomalies, respectively. The two types of anomalies cancel each other out if and only if the Chan–Paton gauge group is either $SO(32)$ or $E_8 \times E_8$ [Green and Schwarz (1984)]. Type II supergravity is anomaly-free [Alvarez-Gaumé and Witten (1984)].

Exercise 5.1. *Prove the following ten dimensional Dirac matrix identities (square brackets mean to antisymmetrize over the corresponding set of indices):*
(a)

$$\Gamma^{11}\,\Gamma^{[\mu_1}\cdots\Gamma^{\mu_n]} = \frac{(-1)^{\left[\frac{n}{2}\right]}\,n!}{\left[(10-n)!\right]^2}\,\epsilon^{\mu_1\cdots\mu_n}{}_{\nu_1\cdots\nu_{10-n}}\,\Gamma^{[\nu_1}\cdots\Gamma^{\nu_{10-n}]}\ ,$$

$$\Gamma^{[\mu_1}\cdots\Gamma^{\mu_n]}\,\Gamma^{11} = \frac{(-1)^{\left[\frac{n+1}{2}\right]}\,n!}{\left[(10-n)!\right]^2}\,\epsilon^{\mu_1\cdots\mu_n}{}_{\nu_1\cdots\nu_{10-n}}\,\Gamma^{[\nu_1}\cdots\Gamma^{\nu_{10-n}]}\ .$$

(b)

$$\Gamma^{\nu}\,\Gamma^{[\mu_1}\cdots\Gamma^{\mu_n]} = \frac{\Gamma^{[\nu}\,\Gamma^{\mu_1}\cdots\Gamma^{\mu_n]}}{n+1} + \frac{n}{(n-1)!}\,\eta^{\nu[\mu_1}\,\Gamma^{\mu_2}\cdots\Gamma^{\mu_n]}\ ,$$

$$\Gamma^{[\mu_1}\cdots\Gamma^{\mu_n]}\,\Gamma^{\nu} = \frac{\Gamma^{[\mu_1}\cdots\Gamma^{\mu_n}\,\Gamma^{\nu]}}{n+1} + \frac{n}{(n-1)!}\,\eta^{\nu[\mu_n}\,\Gamma^{\mu_1}\cdots\Gamma^{\mu_{n-1}]}\ .$$

We will examine the R–R sectors *before* the imposition of the physical super-Virasoro constraints on the spectrum. The R-sector spinors are then representations of the spin cover $Spin(10)$ of the ten-dimensonal rotation group and everything will be written in ten dimensional notation. The two irreducible, Majorana–Weyl representations of $Spin(10)$ are the spinor $\mathbf{16}_s$ and conjugate spinor $\mathbf{16}_c$. From a group theoretic perspective then, the massless states of the two R–R sectors are characterized by their Clebsch–Gordan decompositions:
Type IIA :

$$\mathbf{16}_s \otimes \mathbf{16}_c = [0] \oplus [2] \oplus [4]\ . \tag{5.5}$$

Type IIB :

$$\mathbf{16}_s \otimes \mathbf{16}_s = [1] \oplus [3] \oplus [5]_+\ . \tag{5.6}$$

Here $[n]$ denotes the irreducible n-times antisymmetrized representation of $Spin(10)$, which from a field theoretic perspective corresponds to a completely antisymmetric tensor of rank n, or in other words an "n-form". The $+$ subscript indicates a certain self-duality condition which will be discussed below. The n-forms arise by explicitly decomposing the Ramond bi-spinor $|\psi_{\rm l}\rangle_{\rm R} \otimes |\psi_{\rm r}\rangle_{\rm R}$ in a basis of antisymmetrized Dirac matrices:

Type IIA :

$$(\psi_{\rm l})_a\,(\psi_{\rm r})_{\bar b} = \sum_{n\,\rm even} F^{(n)}_{\mu_1\cdots\mu_n}\left(\Gamma^{[\mu_1}\cdots\Gamma^{\mu_n]}\right)_{a\bar b}\ . \tag{5.7}$$

Type IIB :

$$(\psi_{\rm l})_a\,(\psi_{\rm r})_{b} = \sum_{n\,\rm odd} F^{(n)}_{\mu_1\cdots\mu_n}\left(\Gamma^{[\mu_1}\cdots\Gamma^{\mu_n]}\right)_{ab}\ . \tag{5.8}$$

Inverting these relations gives the n-forms explicitly as

$$F^{(n)}_{\mu_1\cdots\mu_n} = {}_{\rm R}\langle\overline{\psi}_{\rm l}|\,\Gamma_{[\mu_1}\cdots\Gamma_{\mu_n]}|\psi_{\rm r}\rangle_{\rm R}\ . \tag{5.9}$$

Because of the GSO projection, the states $|\psi_{\rm l,r}\rangle_{\rm R}$ have definite Γ^{11} eigenvalue ± 1. The Γ-matrix relations of Exercise 5.1 (a) thereby generate an isomorphism

$$F^{(n)}_{\mu_1\cdots\mu_n} \sim \epsilon_{\mu_1\cdots\mu_n}{}^{\nu_1\cdots\nu_{10-n}}\,F^{(10-n)}_{\nu_1\cdots\nu_{10-n}}\ . \tag{5.10}$$

This identifies the irreducible representations $[n] \cong [10-n]$, and in particular $[5]_+$ is "self-dual".

Exercise 5.2. **(a)** *Verify that the number of independent components of the antisymmetric tensor fields* $F^{(n)}_{\mu_1\cdots\mu_n}$ *agrees with that of the tensor product of two Majorana–Weyl spinors in ten dimensions.*
(b) *Show that the Dirac–Ramond equations for* $|\psi_{\rm l}\rangle_{\rm R}$ *and* $|\psi_{\rm r}\rangle_{\rm R}$ *are equivalent to the field equations*

$$\partial_{[\mu}F^{(n)}_{\mu_1\cdots\mu_n]} = 0\ , \quad \partial^\mu F^{(n)}_{\mu\mu_2\cdots\mu_n} = 0\ .$$

From Exercise 5.2 (b) it follows that the physical constraints in the R–R sector are equivalent to the "Maxwell equation of motion" and "Bianchi identity" for an antisymmetric tensor field strength:

$$\boxed{F^{(n)}_{\mu_1\cdots\mu_n} = \partial_{[\mu_1}C^{(n-1)}_{\mu_2\cdots\mu_n]}}\ , \tag{5.11}$$

generalizing the familiar equations of motion of electrodynamics. The n-forms $F^{(n)}_{\mu_1\cdots\mu_n}$ are called "Ramond–Ramond fields", while $C^{(n)}_{\mu_1\cdots\mu_n}$ are called "Ramond–Ramond potentials". The isomorphism (5.10) corresponds to an electric-magnetic duality which exchanges equations of motion and Bianchi

identities. It relates the fields $C^{(n)}$ and $C^{(8-n)}$, which are thereby treated on equal footing in string theory. Using (5.5)–(5.8) we can look at the totality of Ramond–Ramond potentials in the Type II theories:
<u>Type IIA :</u>

$$C^{(1)} \qquad C^{(3)} \qquad C^{(5)} \qquad C^{(7)} . \tag{5.12}$$

Here the fields $C^{(5)}$ and $C^{(7)}$ are dual to $C^{(3)}$ and $C^{(1)}$, respectively.
<u>Type IIB :</u>

$$C^{(0)} \qquad C^{(2)} \qquad C^{(4)} \qquad C^{(6)} \qquad C^{(8)} . \tag{5.13}$$

Here the field $C^{(4)}$ is self-dual, while $C^{(6)}$ and $C^{(8)}$ are dual to $C^{(2)}$ and $C^{(0)}$, respectively.

5.1.3 *Ramond–Ramond Charges*

We finally come to the crux of the matter here. Recall that the antisymmetric NS–NS sector tensor $B_{\mu\nu}$ couples directly to the string worldsheet, because as we saw in (3.36) its vertex operator couples directly to $B_{\mu\nu}$. In other words, the string carries (electric) "charge" with respect to $B_{\mu\nu}$. However, the situation for the R–R potentials $C^{(n)}$ above is very different, because from (5.7), (5.8) it follows that the vertex operators for the R–R states involve only the R–R *fields* $F^{(n+1)}$ and so only the field strengths, and not potentials, couple to the string. Thus elementary, perturbative string states cannot carry any charge with respect to the R–R gauge fields $C^{(p+1)}$.

We are thereby forced to search for non-perturbative degrees of freedom which couple to these potentials. Clearly, these objects must be "p-branes", which are defined to be p-dimensional extended degrees of freedom that sweep out a $p+1$-dimensional "worldvolume" as they propagate in time, generalizing the notion of a string. The minimal coupling would then involve the potential $C^{(p+1)}$ multiplied by the induced volume element on the hypersurface, and it takes the form

$$q \int \mathrm{d}^{p+1}\xi \; \epsilon^{a_0\cdots a_p} \, \frac{\partial x^{\mu_1}}{\partial \xi^{a_0}} \cdots \frac{\partial x^{\mu_{p+1}}}{\partial \xi^{a_p}} \; C^{(p+1)}_{\mu_1\cdots\mu_{p+1}} , \tag{5.14}$$

in complete analogy with the electromagnetic coupling (3.35) and the B-field minimal coupling (3.36). Gauge transformations shift $C^{(p+1)}$ by p-forms $\Lambda^{(p)}$ according to

$$C^{(p+1)}_{\mu_1\cdots\mu_{p+1}} \longmapsto C^{(p+1)}_{\mu_1\cdots\mu_{p+1}} + \partial_{[\mu_1} \Lambda^{(p)}_{\mu_2\cdots\mu_{p+1}]} , \tag{5.15}$$

and they leave the field strength $F^{(p+2)}$ and all physical string states invariant. As usual in quantum field theory, from (5.14) it follows that they also leave the quantum theory invariant if at the same time the p-brane wavefunction Ψ transforms as

$$\Psi \longmapsto e^{\,i\,q\int_{p-\mathrm{brane}}\Lambda^{(p)}}\,\Psi\ . \tag{5.16}$$

This defines the "Ramond–Ramond charge" q of the p-brane with respect to the "gauge field" $C^{(p+1)}$.

The upshot of this analysis is that, since perturbative string states are R–R neutral, string theory has to be complemented with *non-perturbative* states which carry the R–R charges. These objects are known as "Dirichlet p-branes", or "D-branes" for short, and they are dynamical objects which are extended in p spatial dimensions. In the remainder of this chapter we will present a systematic derivation of how these states arise in string theory.

5.2 T-Duality for Closed Strings

To systematically demonstrate the existence of D-branes in superstring theory, i.e. to quantify their origin and explain what they are, we will first need to describe a very important string theoretical symmetry. It is a crucial consequence of the *extended* nature of strings and it has no field theory analog. We will first consider closed bosonic strings with worldsheet the cylinder and coordinates $-\infty < \tau < \infty$, $0 \leq \sigma < 2\pi$.

Let us begin by recalling the mode expansions of the string embedding fields:

$$\begin{aligned} x^\mu(\tau,\sigma) &= x^\mu_{\mathrm{L}}(\tau+\sigma) + x^\mu_{\mathrm{R}}(\tau-\sigma) \\ &= x^\mu_0 + \tilde{x}^\mu_0 + \sqrt{\frac{\alpha'}{2}}\left(\alpha^\mu_0 + \tilde{\alpha}^\mu_0\right)\tau + \sqrt{\frac{\alpha'}{2}}\left(\alpha^\mu_0 - \tilde{\alpha}^\mu_0\right)\sigma \\ &\quad + (\text{oscillators})\ , \end{aligned} \tag{5.17}$$

where we focus our attention on only the zero modes as the oscillator contributions will play no role in the present discussion. The total (center of mass) spacetime momentum of the string is given by integrating the τ derivative of (5.17) over σ to get

$$p^\mu_0 = \frac{1}{\sqrt{2\alpha'}}\left(\alpha^\mu_0 + \tilde{\alpha}^\mu_0\right)\ . \tag{5.18}$$

Since the oscillator terms are periodic, a periodic shift along the string changes the function (5.17) as

$$x^\mu(\tau,\sigma+2\pi) = x^\mu(\tau,\sigma) + 2\pi\sqrt{\frac{\alpha'}{2}}\left(\alpha_0^\mu - \tilde{\alpha}_0^\mu\right) , \tag{5.19}$$

and so requiring that the embedding be single-valued under $\sigma \mapsto \sigma + 2\pi$ yields the constraint

$$\alpha_0^\mu = \tilde{\alpha}_0^\mu = \sqrt{\frac{\alpha'}{2}}\, p_0^\mu \tag{5.20}$$

with p_0^μ real. This is of course just what we worked out before in Section 2.3.

All of this is true if the spatial directions in spacetime all have infinite extent. We would now like to see how the mode expansion is modified if one of the directions, say x^9, is compact. For this, we "compactify" x^9 on a circle $\mathbf{S}^1$ of radius R. This means that the spacetime coordinate is periodically identified as

$$x^9 \sim x^9 + 2\pi R\ . \tag{5.21}$$

Then the basis wavefunction $\mathrm{e}^{\,\mathrm{i}\, p_0^9 x^9}$, which is also the generator of translations in the x^9 direction, is single-valued under (5.21) *only* if the momentum p_0^9 is quantized according to

$$p_0^9 = \frac{n}{R} \tag{5.22}$$

for some integer n. This quantization condition applies to any quantum system and is not particular to strings. Using (5.18) we then have the zero mode constraint

$$\alpha_0^9 + \tilde{\alpha}_0^9 = \frac{2n}{R}\sqrt{\frac{\alpha'}{2}}\ . \tag{5.23}$$

In addition, under a periodic shift $\sigma \mapsto \sigma + 2\pi$ along the string, the string can "wind" around the spacetime circle. This means that, since by (5.21) the coordinate x^9 is no longer a periodic function, the relation

$$x^9(\tau,\sigma+2\pi) = x^9(\tau,\sigma) + 2\pi w R \tag{5.24}$$

is an allowed transformation for any integer w. For a fixed "winding number" $w \in \mathbb{Z}$, this is represented by adding the term $wR\sigma$ to the mode expansion (5.17) for $x^9(\tau,\sigma)$, and comparing with (5.19) then yields a second zero mode constraint

$$\alpha_0^9 - \tilde{\alpha}_0^9 = wR\sqrt{\frac{2}{\alpha'}}\ . \tag{5.25}$$

Solving (5.23) and (5.25) simultaneously then gives

$$\begin{aligned} \alpha_0^9 &= p_{\rm L}\sqrt{\frac{\alpha'}{2}} \ , \quad p_{\rm L} = \frac{n}{R} + \frac{wR}{\alpha'} \ , \\ \tilde\alpha_0^9 &= p_{\rm R}\sqrt{\frac{\alpha'}{2}} \ , \quad p_{\rm R} = \frac{n}{R} - \frac{wR}{\alpha'} \ , \end{aligned} \tag{5.26}$$

where $p_{\rm L}$ and $p_{\rm R}$ are called the "left-moving and right-moving momenta".

Let us now consider the mass spectrum in the remaining uncompactified 1+8 dimensions, which is given by

$$\begin{aligned} m^2 &= -p_\mu\,p^\mu \ , \quad \mu = 0,1,\ldots,8 \\ &= \frac{2}{\alpha'}\left(\alpha_0^9\right)^2 + \frac{4}{\alpha'}\left(N-1\right) \\ &= \frac{2}{\alpha'}\left(\tilde\alpha_0^9\right)^2 + \frac{4}{\alpha'}\left(\tilde N-1\right) \ , \end{aligned} \tag{5.27}$$

where the second and third equalities come from the $L_0 = 1$ and $\tilde L_0 = 1$ constraints, respectively. The sum and difference of these left-moving and right-moving Virasoro constraints give, respectively, the Hamiltonian and level-matching formulas. Here we find that they are modified from those of Section 3.2 to

$$\begin{aligned} & m^2 = \frac{n^2}{R^2} + \frac{w^2R^2}{\alpha'^2} + \frac{2}{\alpha'}\left(N+\tilde N-2\right) \ , \\ & nw + N - \tilde N = 0 \ . \end{aligned} \tag{5.28}$$

We see that there are extra terms present in both the mass formula and the level-matching condition, in addition to the usual oscillator contributions, which come from the "Kaluza–Klein tower of momentum states" $n \in \mathbb{Z}$ and the "tower of winding states" $w \in \mathbb{Z}$ of the string. The winding modes are a purely *stringy* phenomenon, because only a string can wrap non-trivially around a circle. The usual non-compact states are obtained by setting $n = w = 0$, and, in particular, the massless states arise from taking $n = w = 0$ and $N = \tilde N = 1$.

It is interesting to compare the large and small radius limits of the compactified string theory:

<u>$R \to \infty$:</u> In this limit all $w \neq 0$ winding states disappear as they are energetically unfavourable. But the $w = 0$ states with all values of $n \in \mathbb{Z}$

go over to the usual continuum of momentum zero modes k^μ in the mass formula. This is the same thing that would happen in quantum field theory. In this limit, the spacetime momenta are related by $p_{\rm R} = p_{\rm L}$.
$\underline{R \to 0:}$ In this limit all $n \neq 0$ momentum states become infinitely massive and decouple. But now the pure winding states $n = 0$, $w \neq 0$ form a continuum as it costs very little energy to wind around a small circle. So as $R \to 0$ an extra uncompactified dimension reappears. This is in marked contrast to what would occur in quantum field theory, whereby all surviving fields in the limit would be just independent of the compact coordinate x^9. In this limit the spacetime momenta are related by $p_{\rm R} = -p_{\rm L}$.

The appearence of the extra dimension in the small radius limit requires an explanation. This stringy behaviour is the earmark of "T-duality", as we shall now explain. The mass formula (5.28) for the spectrum is invariant under the simultaneous exchanges

$$\boxed{n \longleftrightarrow w \ , \quad R \longleftrightarrow R' = \frac{\alpha'}{R} \ ,} \tag{5.29}$$

which by (5.26) is equivalent to the zero-mode transformations

$$\alpha_0^9 \longmapsto \alpha_0^9 \ , \quad \tilde{\alpha}_0^9 \longmapsto -\tilde{\alpha}_0^9 \ . \tag{5.30}$$

This symmetry of the compactified string theory is known as a "T-duality symmetry" [Giveon, Porrati and Rabinovici (1994)]. The string theory compactified on the circle of radius R' (with momenta and windings interchanged) is called the "T-dual string theory". The process of going from a compactified string theory to its dual is called "T-dualization".

We can also realize this symmetry at the full interacting level of massive states by a "spacetime parity transformation" of the worldsheet right-movers as

$$\boxed{{\rm T} \ : \ x_{\rm L}^9(\tau+\sigma) \longmapsto x_{\rm L}^9(\tau+\sigma) \ , \quad x_{\rm R}^9(\tau-\sigma) \longmapsto -x_{\rm R}^9(\tau-\sigma) \ .} \tag{5.31}$$

This is called a "T-duality transformation". It holds because the entire worldsheet quantum field theory is invariant under the rewriting of the radius R theory in terms of the dual string coordinates

$$x'^9 = x_{\rm L}^9 - x_{\rm R}^9 \ . \tag{5.32}$$

It is simply a matter of convention whether to add or subtract the two worldsheet sectors, since both choices solve the two-dimensional wave equation of Section 2.3. The only change which occurs is that the zero-mode spectrum of the new variable $x'^9(\tau,\sigma)$ is that of the T-dual $R' = \frac{\alpha'}{R}$ string theory. The dual theories are physically equivalent, in the sense that all quantum correlation functions are invariant under the rewriting (5.32). It follows that T-duality $R \leftrightarrow R' = \frac{\alpha'}{R}$ is an *exact* quantum symmetry of perturbative closed string theory.

5.2.1 *String Geometry*

One of the many profound consequences of the T-duality symmetry of closed strings is on the nature of spacetime geometry as seen by strings [Amati, Ciafaloni and Veneziano (1989); Gross and Mende (1988); Veneziano (1986)]. In particular, the "moduli space" of these compactified string theories, which in this simple case is the space of circle radii R parametrizing the fields, is the semi-infinite line $R \geq \ell_s = \sqrt{\alpha'}$, rather than the classical bound $R \geq 0$ (Fig. 5.1). By T-duality, very small circles are equivalent to very large ones in string theory. Thus strings see spacetime geometry very differently than ordinary point particles do. In this sense, strings modify classical general relativity at very short distance scales.

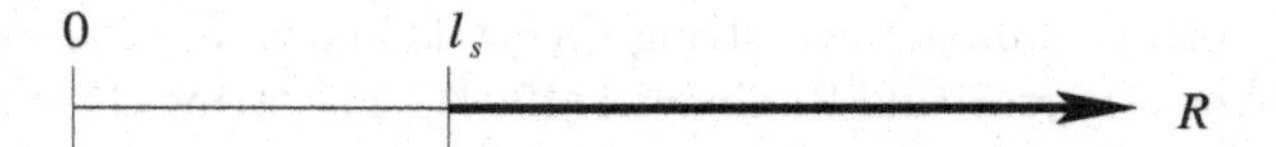

Fig. 5.1 The moduli space of string theory compactified on a circle of radius R. The circles seen by string theory correspond to the line $R \geq \ell_s$, with ℓ_s the fixed point $R' = R$ of the T-duality transformation $R \mapsto \frac{\alpha'}{R}$. Closed string probes cannot see very small radius circles with $0 \leq R < \ell_s$ due to the finite intrinsic size ℓ_s of the string. The entire quantum string theory on a circle of radius $R \in [0, \ell_s)$ can be mapped onto a completely equivalent one with $R \geq \ell_s$ by using T-duality.

5.3 T-Duality for Open Strings

We will now turn to the case of bosonic open strings with worldsheet the infinite strip $-\infty < \tau < \infty$, $0 \leq \sigma \leq \pi$. Open strings do not wind around the periodic direction of spacetime, and so they have no quantum number comparable to w. Thus something very different must happen as compared to the closed string case. The absence of w in fact means that the open string theory looks more like a quantum field theory in the limit $R \to 0$, in that states with non-zero Kaluza–Klein momentum $n \neq 0$ become infinitely massive but no new continuum states arise. So we have reached an apparent paradox. Unitarity requires any fully consistent string theory to possess both open and closed strings. The reason for this is another worldsheet quantum duality which we will discuss briefly in Section 7.3. However, the open strings effectively live in *nine* spacetime dimensions as $R \to 0$, while the closed strings live in *ten* dimensions. The way out of this paradox is to note that the interior of the open string still vibrates in ten dimensions, because there the theory actually resembles that of a closed string. The distinguished parts are the string *endpoints* which are restricted to lie on a nine dimensional hyperplane in spacetime.

To quantify these statements, we recall the open string mode expansions

$$x^\mu(\tau\pm\sigma) = \frac{x_0^\mu}{2} \pm \frac{x_0'^\mu}{2} + \sqrt{\frac{\alpha'}{2}}\,\alpha_0^\mu\,(\tau\pm\sigma) + \mathrm{i}\sqrt{\frac{\alpha'}{2}}\sum_{n\neq 0}\frac{\alpha_n^\mu}{n}\ \mathrm{e}^{-\mathrm{i}n(\tau\pm\sigma)}\ , \tag{5.33}$$

with

$$\alpha_0^\mu = \sqrt{2\alpha'}\,p_0^\mu \tag{5.34}$$

and the total embedding coordinates

$$\begin{aligned} x^\mu(\tau,\sigma) &= x^\mu(\tau+\sigma) + x^\mu(\tau-\sigma) \\ &= x_0^\mu + \alpha' p_0^\mu\,\tau + \mathrm{i}\sqrt{2\alpha'}\sum_{n\neq 0}\frac{\alpha_n^\mu}{n}\ \mathrm{e}^{-\mathrm{i}n\tau}\cos(n\sigma)\ . \end{aligned} \tag{5.35}$$

Note that the arbitrary integration constant $x_0'^\mu$ has dropped out of (5.35) to give the usual open string coordinates that we found in Section 2.3. Again we will put x^9 on a circle of radius R, so that

$$x^9(\tau,\sigma) \sim x^9(\tau,\sigma) + 2\pi R\ , \quad p_0^9 = \frac{n}{R} \tag{5.36}$$

with $n \in \mathbb{Z}$.

To get the T-dual open string coordinate, we use the "doubling trick" of Section 2.3 which may be used to generate the open string mode expansion from the closed one (as in (5.33)). According to what we saw in the previous section, this implies that we should reflect the right-movers, i.e. set $x^9(\tau+\sigma) \mapsto x^9(\tau+\sigma)$ and $x^9(\tau-\sigma) \mapsto -x^9(\tau-\sigma)$. The desired embedding function is therefore given by

$$
\begin{aligned}
x'^9(\tau,\sigma) &= x^9(\tau+\sigma) - x^9(\tau-\sigma) \\
&= x_0'^9 + 2\alpha' \frac{n}{R}\sigma + \sqrt{2\alpha'} \sum_{n\neq 0} \frac{\alpha_n^9}{n}\, \mathrm{e}^{-in\tau} \sin(n\sigma) \ . \qquad (5.37)
\end{aligned}
$$

Notice that the zero mode sector of (5.37) is independent of the worldsheet time coordinate τ, and hence the new string embedding carries no momentum. Thus the dual string is *fixed*. Since $\sin(n\sigma) = 0$ at $\sigma = 0, \pi$, the endpoints do not move in the x^9 direction, i.e. $\partial_\tau x'^9|_{\sigma=0,\pi} = 0$. That is to say, instead of the usual Neumann boundary condition

$$
\partial_\perp x^9 \Big|_{\sigma=0,\pi} \equiv \partial_\sigma x^9 \Big|_{\sigma=0,\pi} = 0 \qquad (5.38)
$$

with $\partial_\perp$ the "normal derivative" to the boundary of the string worldsheet, we now have

$$
\partial_\parallel x'^9 \Big|_{\sigma=0,\pi} \equiv \partial_\tau x'^9 \Big|_{\sigma=0,\pi} = 0 \qquad (5.39)
$$

with $\partial_\parallel$ the "tangential derivative" to the worldsheet boundary. This gives the "Dirichlet boundary condition" that the open string endpoints are at a fixed place in spacetime given by the formula

$$
\boxed{x'^9(\tau,\pi) - x'^9(\tau,0) = \frac{2\pi\alpha'\, n}{R} = 2\pi n R' \ .} \qquad (5.40)
$$

Thus the endpoints $x'^9|_{\sigma=0,\pi}$ are equal up to the periodicity of the T-dual dimension. We may thereby regard this formula as defining an "open string of winding number $n \in \mathbb{Z}$" (Fig. 5.2).

The open string ends are still free to move in the other 1+8 directions that are not T-dualized, which constitute a hyperplane called a "D-brane" [Dai, Leigh and Polchinski (1989); Hořava (1989)]. As there are eight spatial dimensions, we call it more specifically a "D8-brane". Generally, T-dualizing m directions of the spacetime gives Dirichlet boundary conditions in the m directions, and hence a hyperplane with $p = 9 - m$ spatial dimensions which we will call a "Dp-brane". We conclude that T-duality, as a symmetry of the fully consistent string theory (containing both open and closed strings), necessitates *D-branes.*

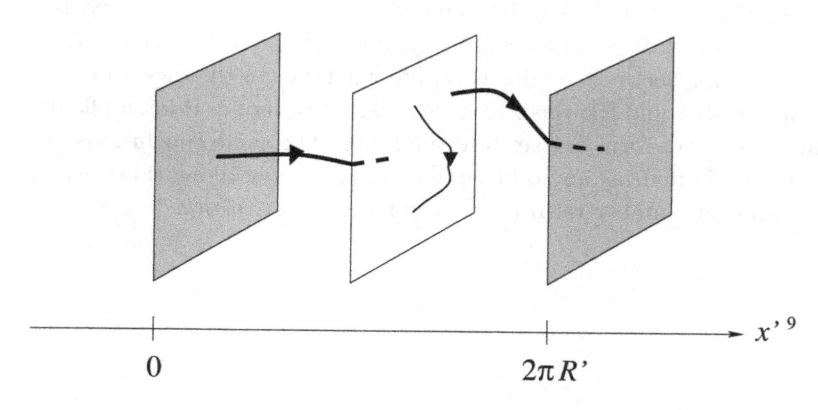

Fig. 5.2 The ends of the T-dual open string coordinate x'^9 attach to hyperplanes in spacetime. The shaded hyperplanes are periodically identified. The thick string has winding number 1, while the thin string has winding number 0.

Exercise 5.3. *Show explicitly that T-duality interchanges the definitions of normal and tangential derivatives, and hence it exchanges Neumann and Dirichlet boundary conditions.*

5.4 T-Duality for Type II Superstrings

Finally, we will now generalize the results of the previous two sections to superstrings. Let us consider the effects of T-duality on the closed, oriented Type II theories. As we have seen, as a right-handed parity transformation, it flips the sign of the right-mover $x^9_{\rm R}(\tau-\sigma)$. By worldsheet supersymmetry, it must do the same on the right-moving fermion fields $\psi^9_-(\tau-\sigma)$, so that

$$\mathrm{T}\,:\,\psi^9_-\,\longmapsto\,-\,\psi^9_-\,. \tag{5.41}$$

This implies that the zero-mode of ψ^9_- in the Ramond sector, which acts as the Dirac matrix Γ^9 on right-movers, changes sign, and hence $\mathrm{T}:\Gamma^{11}\mapsto -\Gamma^{11}$. Thus the relative chirality between left-movers and right-movers is flipped, i.e. T-duality reverses the sign of the GSO projection on right-movers:

$$\mathrm{T}\,:\,P^{\pm}_{\rm GSO}\,\longmapsto\,P^{\mp}_{\rm GSO}\,. \tag{5.42}$$

We conclude that T-duality interchanges the Type IIA and Type IIB superstring theories. It is only a symmetry of the closed string sector, since in the open string sector it relates two different types of theories. Furthermore, since the IIA and IIB theories have different Ramond–Ramond fields, T-duality must transform one set into the other. The same conclusions are reached if one T-dualizes any odd number of spacetime dimensions, while dualizing an even number returns the original Type II theory.

Chapter 6

D-Branes and Gauge Theory

In this chapter we will start working towards a systematic description of D-branes, which were introduced in the previous chapter. We will begin with a heuristic, qualitative description of D-branes, drawing from the way they were introduced in the previous chapter, and painting the picture for the way that they will be analysed. The underlying theme, as we will see, is that they are intimately tied to gauge theory. Indeed, they provide a means of embedding gauge theories into superstring theory. In particular, we will see how to describe their collective coordinates in terms of standard gauge theory Wilson lines. We will then take our first step to describing the low-energy dynamics of D-branes, which will be covered in more detail in the next and final chapter. Here we shall derive the celebrated Born–Infeld action, which will also introduce some further important computational tools.

6.1 D-Branes

Let us begin with a heuristic description of the new extended degrees of freedom that we have discovered in the previous chapter. By a "Dp-brane" we will mean a $p+1$ dimensional hypersurface in spacetime onto which open strings can attach (Fig. 6.1). Such objects arise when we choose *Dirichlet* rather than Neumann boundary conditions for the open strings. More precisely, the Dp-brane is specified by choosing Neumann boundary conditions in the directions along the hypersurface,

$$\partial_\sigma x^\mu \Big|_{\sigma=0,\pi} = 0 \ , \quad \mu = 0,1,\dots,p \ , \tag{6.1}$$

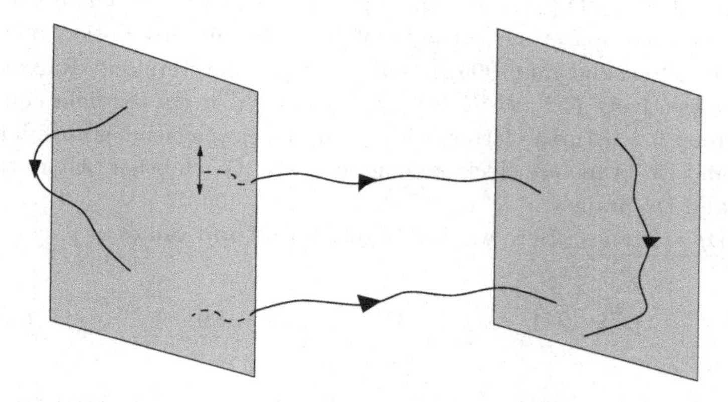

Fig. 6.1 A pair of D-branes (shaded regions) with open strings (wavy lines) attached (with Dirichlet boundary conditions). The string ends are free to move along the hyperplanes. The corresponding open string coordinates satisfy Neumann boundary conditions in the directions along the D-branes and Dirichlet boundary conditions in the directions transverse to the D-branes.

and Dirichlet boundary conditions in the transverse directions,

$$\delta x^\mu \Big|_{\sigma=0,\pi} = 0 \ , \quad \mu = p+1, \ldots, 9 \ . \tag{6.2}$$

In string perturbation theory, the position of the Dp-brane is fixed at the boundary coordinates $x^{p+1}, \ldots, x^9$ in spacetime, corresponding to a particular string theory background (i.e. a solution to the classical string equations of motion), while $x^0, \ldots, x^p$ are free to move along the $p+1$ dimensional hypersurface. Using R–R vertex operators, it can be shown that they must couple Dp-branes and R–R potentials $C^{(p+1)}$ [Polchinski (1995)], and furthermore that fundamental (perturbative) strings cannot carry the R–R charges. Given this fact, we can readily deduce from (5.12) and (5.13) what sorts of D-branes live in the different Type II superstring theories:

Type IIA Dp – Branes : These branes exist for all even values of p,

$$p = 0 \quad 2 \quad 4 \quad 6 \quad 8 \ . \tag{6.3}$$

The case $p = 0$ is a "D-particle", while $p = 8$ describes a "domain wall" in ten dimensional spacetime (in light of the solitonic description of D-branes [Duff, Khuri and Lu (1995)]). The corresponding Ramond–Ramond fields are respectively $F^{(2)}$, $F^{(4)}$, $F^{(6)}$, $F^{(8)}$, and $F^{(10)}$. By the field equations (Exercise 6.2 (b)), the latter field admits no propagating states. The D0-brane and D6-brane are electromagnetic duals of each other, as are the D2-brane and D4-brane.

<u>Type IIB Dp – Branes :</u> Here we find branes for all odd values of p,

$$p = \quad -1 \qquad 1 \qquad 3 \qquad 5 \qquad 7 \qquad 9\,. \tag{6.4}$$

The case $p = -1$ describes an object which is localized in time and corresponds to a "D-instanton", while $p = 1$ is a "D-string". The D9-branes are spacetime filling branes, with no coupling to any R–R field strength, while $p = 3$ yields the self-dual D3-brane (which we note has worldvolume of observable 3+1-dimensions). The D-instanton and D7-brane are electromagnetic duals of one another, as are the D1-brane and the D5-brane.

In what follows though we will work towards a *non-perturbative* description of D-branes. We will find that the massless modes of open strings are associated with the fluctuation modes of the D-branes themselves, so that, non-perturbatively, D-branes become dynamical p-dimensional objects. Let us give a heuristic description of how the dynamical degrees of freedom of a D-brane arise from the massless string spectrum in a *fixed* D-brane background. For this, we recall that the open string spectrum contains a massless $SO(8)$ vector A_μ. In the presence of the D-brane, the gauge field $A_\mu(x)$ decomposes into components parallel and perpendicular to the D-brane worldvolume. Because the endpoints of the strings are tied to the worldvolume, these massless fields can be interpreted in terms of a low-energy field theory on the D-brane worldvolume. Precisely, a ten-dimensional gauge field $A_\mu(x)$, $\mu = 0, 1, \ldots, 9$ will split into components A_a, $a = 0, 1, \ldots, p$ corresponding to a $U(1)$ gauge field on the Dp-brane, and components Φ^m, $m = p + 1, \ldots, 9$ which are scalar fields describing the fluctuations of the Dp-brane in the $9-p$ transverse directions. We will find that the low-energy dynamics of such a configuration is governed by a supersymmetric gauge theory, obtained from the dimensional reduction to the Dp-brane of maximally supersymmetric Yang–Mills theory in ten spacetime dimensions. In this chapter and the next we will work towards making these statements precise and explicit.

6.2 Wilson Lines

To make the discussion of the previous section more quantitative, we will first need an elementary, but perhaps not so widely appreciated, result from quantum mechanics, which we leave as an exercise.[1]

Exercise 6.1. *A relativistic particle of mass m and charge q in d Euclidean spacetime dimensions propagates in a background electromagnetic vector potential $A_\mu(x)$ according to the action*

$$S = \int \mathrm{d}\tau \, \left(\frac{m}{2}\,\dot{x}^\mu\,\dot{x}_\mu - \mathrm{i}\,q\,\dot{x}^\mu\,A_\mu\right) .$$

Show that if the d-th direction is compactified on a circle of radius R, then a constant gauge field $A_\mu = -\delta_{\mu,d}\,\frac{\theta}{2\pi R}$ induces a fractional canonical momentum

$$p^d = \frac{n}{R} + \frac{q\,\theta}{2\pi R} .$$

We will begin by again compactifying the x^9 direction of spacetime, $x^9 \sim x^9 + 2\pi R$, and introduce $U(N)$ Chan–Paton factors for a Type II oriented open string. Consider the open string vertex operator corresponding to a *constant*, background abelian gauge field

$$A_\mu = \delta_{\mu,9}\begin{pmatrix} \dfrac{\theta_1}{2\pi R} & & 0 \\ & \ddots & \\ 0 & & \dfrac{\theta_N}{2\pi R} \end{pmatrix} , \tag{6.5}$$

where θ_i, $i = 1, \dots, N$ are constants. The introduction of this electromagnetic background generically breaks the Chan–Paton gauge symmetry as $U(N) \to U(1)^N$, as (6.5) is clearly only invariant under an abelian subgroup of $U(N)$. *Locally*, it is a trivial, pure gauge configuration, since its non-vanishing component can be written as

$$A_9 = -\mathrm{i}\,\Lambda^{-1}\,\partial_9\Lambda \ , \quad \Lambda(x) = \begin{pmatrix} \mathrm{e}^{\,\mathrm{i}\,\theta_1 x^9/2\pi R} & & 0 \\ & \ddots & \\ 0 & & \mathrm{e}^{\,\mathrm{i}\,\theta_N x^9/2\pi R} \end{pmatrix} . \tag{6.6}$$

[1]The free kinetic term in the action below is not the same as the one we studied in Section 2.1, but is rather a point particle version of the Polyakov action. The second term is the usual minimal coupling of a particle to a gauge potential.

This means that we can gauge A_9 away by a local gauge transformation. But this is not true *globally*, because the compactness of the x^9 direction still leads to non-trivial effects involving the background (6.5). This is manifested in the fact that the gauge transformation $\Lambda(x)$ of (6.6) is not a single-valued function on spacetime, but has the singular behaviour

$$\Lambda(x^9 + 2\pi R) = W \cdot \Lambda(x) \ , \quad W = \begin{pmatrix} e^{\,i\,\theta_1} & & 0 \\ & \ddots & \\ 0 & & e^{\,i\,\theta_N} \end{pmatrix} . \tag{6.7}$$

All charged (gauge invariant) states pick up the phase factor W under a periodic translation $x^9 \mapsto x^9 + 2\pi R$ due to the trivializing gauge transformation, and the effects of the trivial gauge field (6.5) are still felt by the theory. Thus the configuration A_μ yields no contribution to the (local) equations of motion, and its only effects are in the holonomies as the string ends wind around the compactified spacetime direction.

The phase factor W in (6.7) is in fact just the "Wilson line" for the given gauge field configuration (i.e. the group element corresponding to the gauge field), represented by the exponential of the vertex operator (3.42), (3.48) for the open string photon field (6.5):

$$W = \exp\left(i \int dt \ \dot{x}^\mu(t)\, A_\mu\right) = \exp\left(i \int\limits_0^{2\pi R} dx^9 \ A_9\right) . \tag{6.8}$$

This is the observable that appears in exponentiated form in the Polyakov path integral (3.43) for amplitudes, upon introducing the vertex operators compatible with the gauge invariance of the problem. It cannot be set to unity by a gauge transformation and it is ultimately responsible for the symmetry breaking $U(N) \to U(1)^N$. This scenario has a familiar analog in quantum mechanics, the Aharonov–Bohm effect. Placing a solenoid of localized, point-like magnetic flux in a region of space introduces a non-contractible loop, representing charged particle worldlines encircling the source that cannot be contracted to a point because of the singularity introduced by the flux. Although everywhere outside of the source the electromagnetic field is zero, the wavefunctions still acquire a non-trivial phase factor $W \neq 1$ which cannot be removed by the gauge symmetry of the problem. This phase is observable through electron interference patterns obtained from scattering off of the solenoid.

The symmetry breaking mechanism induced by the Wilson line W has a very natural interpretation in the T-dual theory in terms of *D-branes*. From Exercise 6.1 it follows that the string momenta along the x^9 direction are *fractional*, and so the fields in the T-dual description have fractional winding numbers. Recalling the analysis of Section 5.3 (see (5.40)), we see therefore that the two open string endpoints no longer lie on the same hyperplane in spacetime.

To quantify these last remarks, let us consider a Chan–Paton wavefunction $|k;ij\rangle$, as depicted schematically in Fig. 3.4. The state i attached to an end of the open string will acquire a factor $\mathrm{e}^{-\mathrm{i}\,\theta_i x^9/2\pi R}$ due to the gauge transformation (6.6), while state j will have $\mathrm{e}^{\mathrm{i}\,\theta_j x^9/2\pi R}$. The total open string wavefunction will therefore gauge transform to $|k;ij\rangle\cdot\mathrm{e}^{-\mathrm{i}\,(\theta_i-\theta_j)x^9/2\pi R}$, and so under a periodic translation $x^9\mapsto x^9+2\pi R$ it will acquire a Wilson line factor given by

$$|k;ij\rangle \longmapsto \mathrm{e}^{\mathrm{i}\,(\theta_j-\theta_i)}\,|k;ij\rangle\ . \tag{6.9}$$

In Fourier space this should be manifested through spacetime translations in the plane waves $\mathrm{e}^{\mathrm{i}\,p_9 x^9}$, which means that the momentum of the state $|k;ij\rangle$ is given by

$$p_{ij}^9=\frac{n}{R}+\frac{\theta_j-\theta_i}{2\pi R}\ , \tag{6.10}$$

with $n\in\mathbb{Z}$ and $i,j=1,\dots,N$. So by appropriately modifying the mode expansions and the endpoint calculation of Section 5.3, we find that the Dirichlet boundary condition corresponding to this Chan–Paton state is now changed to

$$x'^9(\tau,\pi)-x'^9(\tau,0)=(2\pi n+\theta_j-\theta_i)\ R'\ , \tag{6.11}$$

with $R'=\frac{\alpha'}{R}$ the T-dual compactification radius. Thus, up to an arbitrary additive constant, the open string endpoint in Chan–Paton state i is at the spacetime position

$$\boxed{x_i'^9=\theta_i\,R'=2\pi\alpha'\,(A_9)_{ii}\ ,\quad i=1,\dots,N\ .} \tag{6.12}$$

The locations (6.12) specify N hyperplanes at different positions corresponding to a collection of N parallel, separated D-branes (Fig. 6.2). In particular, from this analysis we may conclude the remarkable fact that T-duality maps gauge fields in open string theory to the localized positions of D-branes in spacetime.

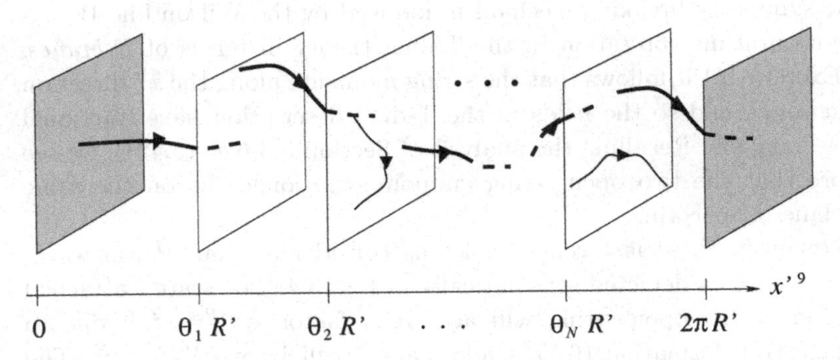

Fig. 6.2 An assembly of N separated, parallel D-branes with open strings attached. The shaded hyperplanes are periodically identified.

6.2.1 *D-Brane Terminology*

We will interpret the original ten-dimensional open strings as lying on N "D9-branes" which fill the spacetime. In this picture the string endpoints can sit anywhere in spacetime and correspond to ordinary Chan–Paton factors. Compactifying $9-p$ coordinates x^m, $m = p+1, \dots, 9$ confines the open string endpoints to N "Dp-brane" hyperplanes of dimension $p+1$. This is a consequence of the corresponding T-duality transformation which maps the Neumann boundary conditions $\partial_\| x^m = 0$ to the Dirichlet ones $\partial_\perp x'^m = 0$, with all other x^a, $a = 0, 1, \dots, p$ still obeying Neumann boundary conditions. Since T-duality interchanges Neumann and Dirichlet boundary conditions (Exercise 5.3), a T-duality transformation along a direction parallel to a Dp-brane produces a D$(p-1)$-brane, while T-duality applied to a direction perpendicular to a Dp-brane yields a D$(p+1)$-brane. Comparing with (6.3) and (6.4), we see that this is in fact the basis of the way that the Type IIA and IIB superstring theories are interchanged under T-duality, as we saw in Section 5.4.

6.3 Collective Coordinates for D-Branes

As we will see very soon, D-branes are in fact *not* rigid hyperplanes in spacetime. They are dynamical, and can fluctuate both in shape and position. For instance, the T-dual theory with D-branes still contains gravity and gauge fields, whose dynamics we will begin describing in the next section.

To see this, let us consider, as we did in (5.27), the 1+8 dimensional massless spectrum, interpreted in the T-dual string theory, for the case where only the coordinate field x^9 is T-dualized. With $\mathcal{N}$ denoting the occupation number of a Chan–Paton state $|k;ij\rangle$, the mass-shell relation reads

$$
\begin{aligned}
m_{ij}^2 &= \left(p_{ij}^9\right)^2 + \frac{1}{\alpha'}\left(\mathcal{N} - 1\right) \\
&= \frac{L_{ij}^2}{(2\pi\alpha')^2} + \frac{1}{\alpha'}\left(\mathcal{N} - 1\right) ,
\end{aligned}
\tag{6.13}
$$

where

$$
L_{ij} = \Big|2\pi n + (\theta_i - \theta_j)\Big|\, R'
\tag{6.14}
$$

is the minimum length of an open string which winds n times between hyperplanes i and j. To examine the massless states, we set $n = 0$ (as it costs energy to wind) and $\mathcal{N} = 1$. Then the string tension T contributes to the energy of a stretched string as

$$
m_{ij}^{(0)} = \frac{R'}{2\pi\alpha'}\Big|\theta_i - \theta_j\Big| = T \cdot L_{ij} \ ,
\tag{6.15}
$$

so that the mass is proportional to the distance L_{ij} between hyperplanes i and j.

Thus, generically massless states *only* arise for non-winding open strings whose ends lie on the same D-brane $i = j$. There are two such types of states/vertex operators that may be characterized as follows:

• $\alpha_{-1}^{\mu}|k;ii\rangle \ , \ V = \partial_{\parallel} x^{\mu}$:

These states correspond to a gauge field $A_\mu(\xi^a)$ on the D-brane with $p+1$ coordinates tangent to the hyperplane, where $\mu, a = 0, 1, \ldots, p$ and $\xi^\mu = x^\mu$ are coordinates on the D-brane worldvolume (in the "static gauge"). It describes the *shape* of the D-brane as a "soliton" background, i.e. as a fixed topological defect in spacetime [Duff, Khuri and Lu (1995)]. The quanta of $A_\mu(\xi^a)$ describe the fluctuations of that background.

• $\alpha_{-1}^{m}|k;ii\rangle \ , \ V = \partial_{\parallel} x^{m} = \partial_{\perp} x'^{\,m}$:

These states correspond to scalar fields $\Phi^m(\xi^a)$, $m = p+1, \ldots, 9$ which originate from the gauge fields in the compact dimensions of the original string theory, and which give the transverse position of the D-brane in the compact dualizing directions. They describe the *shape* of the D-brane as it is embedded in spacetime, analogously to the string embedding coordinates $x^\mu(\tau, \sigma)$.

Thus a flat hyperplane in spacetime has fluctuations described by a certain open string state which corresponds to a *gauge field*. This gives a remarkable description of D-branes in terms of *gauge theory*. These facts are the essence of the gauge theory/geometry correspondences provided by string theory, such as the AdS/CFT correspondence [Aharony *et al* (2000)]. We can in this way describe D-branes using the wealth of experience we have of working with gauge theory, which is the path we shall mostly take in the following. Conversely, it is hoped that D-branes and the string theories in which they live can teach us a lot about gauge theories, such as their strong coupling expansions.

6.3.1 *Non-Abelian Gauge Symmetry*

We have seen above that if none of the D-branes coincide, $\theta_i \neq \theta_j$ for $i \neq j$, then there is a single massless vector state $\alpha_{-1}^M|k;ii\rangle$ associated to each individual D-brane $i = 1, \dots, N$. Together, these states describe a gauge theory with abelian gauge group $U(1)^N$, which is the *generic* unbroken symmetry group of the problem. But suppose now that $k \leq N$ D-branes coincide, say

$$\theta_1 = \theta_2 = \cdots = \theta_k = \theta \ . \tag{6.16}$$

Then from (6.15) it follows that $m_{ij}^{(0)} = 0$ for $1 \leq i, j \leq k$. Thus new massless states appear in the spectrum of the open string theory, because strings which are stretched between these branes can now attain a vanishing length. In all there are k^2 massless vector states which, by the transformation properties of Chan–Paton wavefunctions, form the adjoint representation of a $U(k)$ gauge group. In the *original* open string theory, the coincident position limit (6.16) corresponds to the Wilson line

$$W = \begin{pmatrix} \mathrm{e}^{\,\mathrm{i}\,\theta}\,\mathbb{1}_k & & & 0 \\ & \mathrm{e}^{\,\mathrm{i}\,\theta_{k+1}} & & \\ & & \ddots & \\ 0 & & & \mathrm{e}^{\,\mathrm{i}\,\theta_N} \end{pmatrix} , \tag{6.17}$$

where $\mathbb{1}_k$ denotes the $k \times k$ identity matrix. The background field configuration leaves unbroken a $U(k) \subset U(N)$ subgroup, acting on the upper left $k \times k$ block of (6.17). Thus the D-brane worldvolume now carries a $U(k)$ gauge field $\alpha_{-1}^\mu|k;ij\rangle \leftrightarrow A_\mu(\xi^a)_{ij}$ and, at the same time, a set of k^2 massless scalar fields $\alpha_{-1}^m|k;ij\rangle \leftrightarrow \Phi^m(\xi^a)_{ij}$, where $i, j = 1, \dots, k$.

The geometrical implications of this non-abelian $U(k)$ symmetry are rather profound [Witten (1996)]. The k D-brane positions in spacetime are promoted to a *matrix* $\Phi^m(\xi^a)$ in the adjoint representation of the unbroken $U(k)$ gauge group. This is a curious and exotic feature of D-brane dynamics that is difficult to visualize, and is a consequence of the somewhat surprising aspects of string geometry that were mentioned at the end of Section 5.2. It simply reflects the fact that the T-dual string theory rewrites the $R \ll \ell_s$ limit of the original open string theory in terms of *non-commuting*, matrix-valued spacetime coordinates. These are the variables which are more natural to use as $R \to 0$, and various puzzling features of this limit become clearer in the T-dual picture. This illustrates once again how spacetime geometry is significantly altered by strings at very short distances.

Generally, the gauge symmetry of the theory is broken to the subgroup of $U(N)$ which commutes with the Wilson line W. But if *all* N D-branes coincide, then W belongs to the center of the Chan–Paton gauge group, and we recover the original $U(N)$ gauge symmetry. Since Φ^m is a Hermitian matrix, we can diagonalize it by a $U(N)$ gauge transformation and write

$$\Phi^m = U^m \begin{pmatrix} \phi_1^m & & 0 \\ & \ddots & \\ 0 & & \phi_N^m \end{pmatrix} U^{m\,\dagger} \, . \tag{6.18}$$

The real-valued eigenvalue ϕ_i^m describes the classical position of D-brane i and corresponds to the ground state of the system of N D-branes. The $N \times N$ unitary matrices U^m describe the fluctuations U_{ij}^m, $i \neq j$ about classical spacetime, and they arise from the short open strings connecting D-branes i and j. In this way the off-diagonal elements Φ_{ij}^m, $i \neq j$ may be thought of as "Higgs fields" for the symmetry breaking mechanism [Witten (1996)]. The $U(N)$ symmetry is broken when some (or all) of the D-branes separate, leaving a set of *massive* fields, with mass equal to that of the stretched open strings. We will see precisely how this works dynamically in the next chapter.

6.4 The Born–Infeld Action

We have argued that, associated to any configuration of D-branes, there correspond dynamical gauge fields living on the worldvolumes. At this stage it is natural to ask what sort of gauge theory describes their dynamics. In this section we will begin deriving the appropriate actions which describe the (low-energy) dynamics of D-branes in Type II superstring theory. They will govern the worldvolume dynamics of the gauge fields, the transverse scalar fields, and eventually also the Ramond–Ramond form potentials which couple electrically to the D-brane worldvolume as in (5.14). In this chapter we will consider the really relevant situation that led to the gauge theoretic description above, namely the coupling of free open strings to a background photon field A_μ of constant field strength

$$F_{\mu\nu} = \partial_\mu A_\nu - \partial_\nu A_\mu \ . \tag{6.19}$$

Equivalently, we may regard the open strings as being attached to a spacetime filling D9-brane. The corresponding result for arbitrary dimension D-branes will be derived in the next chapter using T-duality.

We will work at tree-level in open string perturbation theory, and therefore calculate the disc diagram (Fig. 3.5, p. 38). Using conformal invariance we may set the radius of the disc to unity. The complex coordinates on the disc will be written in the polar decomposition

$$z = r \ \mathrm{e}^{\mathrm{i}\,\theta} \ , \tag{6.20}$$

where $0 \leq \theta < 2\pi$ and $0 \leq r \leq 1$.

Exercise 6.2. **(a)** *Show that the solution to the Neumann problem on the disc,*

$$\Delta N(z,z') \equiv \partial_z\,\partial_{\overline{z}} N(z,z') = \delta(z-z') \ ,$$
$$\frac{\partial}{\partial r} N(z,z')\Big|_{r=1} = 0 \ ,$$

is given by

$$N(z,z') = \frac{1}{2\pi}\ln\Big(\big|z-z'\big|\,\big|z-\overline{z}'^{-1}\big|\Big) \ .$$

[Hint: Use a conformal transformation to map the disc onto the upper complex half-plane, and hence apply the method of images.]
(b) *Show that on the boundary of the disc this Green's function can be written as*

$$N(\mathrm{e}^{\mathrm{i}\,\theta},\mathrm{e}^{\mathrm{i}\,\theta'}) = -\frac{1}{\pi}\sum_{n=1}^\infty \frac{\cos\big(n(\theta-\theta')\big)}{n} \ .$$

The effective bosonic string action in the conformal gauge is given by

$$S[x,A] = \frac{1}{4\pi\alpha'}\int \mathrm{d}^2z\ \partial_z x^\mu\,\partial_{\overline{z}} x_\mu - \mathrm{i}\int_0^{2\pi}\mathrm{d}\theta\ \dot{x}^\mu\,A_\mu\Big|_{r=1} \ , \tag{6.21}$$

where

$$\dot{x}^\mu(\theta) \equiv \frac{\partial x^\mu}{\partial\theta} \tag{6.22}$$

and we have used the standard minimal coupling of the point particle at the string endpoint. We are interested in evaluating the gauge-fixed, Euclidean Polyakov path integral

$$Z[F] = \frac{1}{g_s} \int \mathcal{D}x^\mu \ \mathrm{e}^{-S[x,A]} \ , \tag{6.23}$$

where the inverse power of string coupling indicates that we are evaluating a tree-level diagram, and there are no moduli on the disc. By gauge invariance, the result of this functional integration should depend only on the field strength (6.19). Note that the path integral (6.23) can be interpreted as that in (3.43) with the appropriate vertex operator (Wilson line) insertion. We interpret it as the effective gauge field action induced on the D9-brane by integrating out all of the open string modes.[2]

To compute (6.23), we will use the usual "background field gauge" of quantum field theory, in which the string embedding coordinates are expanded as

$$x^\mu = x_0^\mu + \xi^\mu \ , \tag{6.24}$$

where x_0^μ are the constant, worldsheet zero modes of x^μ on the disc. We will also work in the "radial gauge" for the gauge field background,

$$\xi^\mu \, A_\mu(x_0+\xi) = 0 \ , \quad A_\mu(x_0) = 0 \ , \tag{6.25}$$

and with slowly-varying vector potentials,

$$A_\mu(x_0+\xi) = \frac{1}{2}\, F_{\mu\nu}(x_0)\, \xi^\nu + \mathcal{O}(\partial F) \ . \tag{6.26}$$

In other words, we evaluate the path integral (6.23) to leading orders in a derivative expansion in the field strength $F_{\mu\nu}$ (which essentially means that we work with constant $F_{\mu\nu}$).

The path integral measure can be decomposed using (6.24) in terms of bulk and boundary integrations over the disc, i.e. schematically we have

$$\mathcal{D}x^\mu = \prod_{z\in\text{interior}} \mathcal{D}x^\mu(z,\overline{z}) \prod_{\theta\in\text{boundary}} \mathcal{D}\xi^\mu(\theta) \ . \tag{6.27}$$

[2] Note that the Polyakov path integral computes directly the vacuum energy [Abouelsaood *et al* (1988)]. The string partition function is quite different from that of quantum field theory, in that it is more like an S-matrix.

The (Gaussian) bulk integration in the interior of the disc just produces some (functional) normalization factor, corresponding to the closed string sector, which is independent of A_μ. This can be absorbed into an irrelevant normalization of (6.23). Then, integrating out the bulk modes leaves a boundary path integral

$$Z[F] = \frac{1}{g_s} \int \mathrm{d}\vec{x}_0 \int \mathcal{D}\xi^\mu(\theta)\ \mathrm{e}^{-S_{\mathrm{b}}[\xi,A]} , \tag{6.28}$$

where the boundary action is given by

$$S_{\mathrm{b}}[\xi,A] = \frac{1}{2} \int_0^{2\pi} \mathrm{d}\theta \left(\frac{1}{2\pi\alpha'}\, \xi^\mu\, N^{-1}\, \xi_\mu + \mathrm{i}\, F_{\mu\nu}\, \xi^\mu\, \dot{\xi}^\nu \right) \tag{6.29}$$

and N^{-1} is the coordinate space inverse of the boundary Neumann function given in Exercise 6.2. By using the Fourier completeness relation

$$\frac{1}{\pi} \sum_{n=1}^\infty \cos\Big(n(\theta-\theta')\Big) = \delta(\theta-\theta') - \frac{1}{2\pi} \tag{6.30}$$

for $0 \leq \theta, \theta' \leq 2\pi$, and Exercise 6.2 (b), one can easily compute

$$N^{-1}(\theta,\theta') = -\frac{1}{\pi} \sum_{n=1}^\infty n \cos\Big(n(\theta-\theta')\Big) . \tag{6.31}$$

We will expand the non-constant string modes $\xi^\mu(\theta)$ in periodic Fourier series on the circle:

$$\xi^\mu(\theta) = \sum_{n=1}^\infty \Big(a_n^\mu\, \cos(n\theta) + b_n^\mu\, \sin(n\theta) \Big) . \tag{6.32}$$

Then the Feynman measure in (6.28) can be expressed in terms of Fourier modes as

$$\mathcal{D}\xi^\mu(\theta) = \prod_{n=1}^\infty \mathrm{d}a_n^\mu\ \mathrm{d}b_n^\mu . \tag{6.33}$$

We will now use Lorentz-invariance to simplify the form of the action (6.29). The *anti-symmetric* 10×10 matrix $(F_{\mu\nu})$ cannot be diagonalized like a symmetric matrix, but it can be rotated into its canonical

Jordan normal form

$$(F_{\mu\nu}) = \begin{pmatrix} 0 & -f_1 & & & 0 \\ f_1 & 0 & & & \\ & & \ddots & & \\ & & & 0 & -f_5 \\ 0 & & & f_5 & 0 \end{pmatrix} , \qquad (6.34)$$

where the "skew-diagonal" blocks contain the real "skew-eigenvalues" f_l, $l = 1, \ldots, 5$ of $(F_{\mu\nu})$. The path integral (6.28), (6.33) factorizes in this basis into a product of five independent functional Gaussian integrations over the pairs of coordinate modes a_n^{2l-1}, a_n^{2l} and b_n^{2l-1}, b_n^{2l}, where $l = 1, \ldots, 5$. By substituting (6.31) and (6.32) into (6.29), and using standard Fourier properties of the orthogonal trigonometric functions appearing, we find that, for each n and l, the Gaussian Boltzmann weight has rank 2 quadratic form

$$\frac{1}{2} \frac{1}{2\pi n} \frac{1}{2\pi\alpha'} \begin{pmatrix} a_n^{2l-1} , & a_n^{2l} \end{pmatrix} \begin{pmatrix} 1 & -2\pi\alpha' f_l \\ 2\pi\alpha' f_l & 1 \end{pmatrix} \begin{pmatrix} a_n^{2l-1} \\ a_n^{2l} \end{pmatrix} , \qquad (6.35)$$

plus an analogous term for the b_n's. Integrating over each of the a_n's and b_n's thereby yields (up to irrelevant constants)

$$Z[F] = \frac{1}{g_s} \int \mathrm{d}\vec{x}_0 \prod_{l=1}^{5} Z_{2l-1,2l}[f_l] , \qquad (6.36)$$

where

$$Z_{2l-1,2l}[f_l] = \prod_{n=1}^{\infty} \left\{ \left(4\pi^2\alpha' n\right)^2 \left[1 + (2\pi\alpha' f_l)^2\right]^{-1} \right\} \qquad (6.37)$$

is the functional fluctuation determinant arising from the two copies of the Gaussian form (6.35).

To deal with the infinite products in (6.37), we note first of all that $\prod_{n=1}^{\infty} n^2$ diverges. However, it can be regulated by introducing a worldsheet ultraviolet cutoff and thereby absorbing it into an (infinite) renormalization of the string coupling constant g_s. This divergence is due to the *tachyon* mode of the bosonic string, and it originates in the $\theta \to \theta'$ divergence of (6.31). It can therefore be removed by introducing worldsheet supersymmetry, and hence will simply be dropped in what follows. The other infinite product in (6.37) is independent of n and can be evaluated by using "zeta-function regularization" to write

$$\prod_{n=1}^{\infty} c = c^{\zeta(0)} , \tag{6.38}$$

where $\zeta(z)$ is the Riemann zeta-function (3.8). Using (3.9) thereby yields the finite answer

$$Z_{2l-1,2l}[f_l] = \frac{1}{4\pi^2\alpha'} \sqrt{1+(2\pi\alpha' f_l)^2} . \tag{6.39}$$

Finally, we rotate the field strength tensor $F_{\mu\nu}$ back to general form to produce a Lorentz-invariant result. In this way we have found that the partition function (6.23) gives the "open string effective action" [Abouelsaood *et al* (1988); Fradkin and Tseytlin (1985)]

$$\boxed{Z[F] = \frac{1}{(4\pi^2\alpha')^5\, g_s} \int \mathrm{d}\vec{x}_0 \, \sqrt{\det_{\mu,\nu}\left(\delta_{\mu\nu} + 2\pi\alpha' F_{\mu\nu}\right)} ,} \tag{6.40}$$

where we have recalled the origin of the background field dependent terms in (6.39) as the determinant of the quadratic form in (6.35).

The string theoretic action (6.40) is *exact* in α', which is in fact the coupling constant of the original two-dimensional worldsheet field theory defined by (6.21). Thus this result is non-perturbative at the level of the theory on the disc. In other words, the action (6.40) is a truly "stringy" result, containing contributions from *all* massive and massless string states. Remarkably, it actually dates back to 1934, and is known as the "Born–Infeld action" [Born and Infeld (1934)]. This model of non-linear electrodynamics was originally introduced to smoothen out the singular electric field distributions generated by point charges in ordinary Maxwell electrodynamics, thereby yielding a finite total energy. This is quite unlike the situation in Maxwell theory, where the field of a point source is singular at the origin and its energy is infinite. Here the effective distribution of the field has radius of order the string length ℓ_s, and the delta-function singularity is smeared away. It is truly remarkable how string theory captures and revamps this model of non-linear electrodynamics.[3]

[3]In the exercise below, the second equation for the fermionic Green's function defines anti-periodic boundary conditions. Periodic boundary conditions would produce a vanishing functional integral due to the Ramond zero modes.

Exercise 6.3. *In this exercise you will generalize the above derivation to the case of the superstring.*
(a) *Show that by augmenting the bosonic action (6.21) by the fermionic action*

$$S_{\rm ferm}[\psi, A] = \frac{\rm i}{4\pi\alpha'} \int {\rm d}^2 z \, \left(\overline{\psi} \cdot \partial_z \overline{\psi} + \psi \cdot \partial_{\overline{z}} \psi\right) - \frac{\rm i}{2} \int_0^{2\pi} {\rm d}\theta \, \psi^\mu \, F_{\mu\nu} \, \psi^\nu \Big|_{r=1}$$

with $\overline{\psi}^\mu$ and ψ^μ independent complex fermion fields in the bulk of the disc, the total action is invariant under worldsheet supersymmetry transformations.

(b) *Show that the fermionic Green's function on the disc, defined by*

$$\begin{aligned} \partial_z K(z, z') &= \delta(z - z') \ , \\ K({\rm e}^{2\pi\,{\rm i}}\, z, z') \Big|_{r=1} &= -K(z, z') \Big|_{r=1} \ , \end{aligned}$$

can be written on the boundary of the disc as

$$K({\rm e}^{{\rm i}\,\theta}, {\rm e}^{{\rm i}\,\theta'}) = -\frac{1}{\pi} \sum_r \sin\Big(r(\theta - \theta')\Big) \ , \quad r = \frac{1}{2}, \frac{3}{2}, \dots \ .$$

(c) *Show that the superstring path integral yields the same Born–Infeld action (6.40), and at the same time removes the tachyonic divergence. You will need to use the "generalized zeta-function" [Gradshteyn and Ryzhik (1980)]*

$$\zeta(z, a) = \sum_{n=0}^\infty \frac{1}{(n + a)^z}$$

with

$$\zeta(0, 0) = -\frac{1}{2} \ , \quad \zeta(0, 1/2) = 0 \ .$$

Chapter 7

D-Brane Dynamics

In this chapter we will describe various aspects of the dynamics of D-branes in the low-energy limit. We will start from the Born–Infeld action (6.40) and work out its extensions to Dp-branes with $p < 9$, some of their physical properties, and how they couple to the spacetime supergravity fields of the closed string sector. This will then lead us into a description of the dynamics of D-branes in terms of supersymmetric Yang–Mills theory, a more familiar quantum field theory which has sparked the current excitement over the D-brane/gauge theory correspondence, and which unveils some surprising features of D-brane physics. Finally, we will give an elementary calculation of the interaction energy between two separated D-branes, and thereby illustrate the role played by supersymmetry in D-brane dynamics.

7.1 The Dirac–Born–Infeld Action

In the previous chapter we saw that the low-energy dynamics of a D9-brane, induced by the quantum theory of the open strings attached to it, is governed by the Born–Infeld action

$$S_{\rm BI} = \frac{1}{(4\pi^2\alpha')^5\, g_s} \int {\rm d}^{10}x\ \sqrt{-\det_{\mu,\nu}\left(\eta_{\mu\nu} + T^{-1}\, F_{\mu\nu}\right)}\ , \tag{7.1}$$

where we have Wick rotated back to Minkowski signature. Here $T = \frac{1}{2\pi\alpha'}$ is the string tension and $F_{\mu\nu} = \partial_\mu A_\nu - \partial_\nu A_\mu$ is the field strength of the gauge fields living on the D9-brane worldvolume. This low-energy approximation to the full dynamics is good in the static gauge and for slowly-varying field strengths. In general there are derivative corrections from $F_{\mu\nu}$, but (7.1) is nevertheless the *exact* result as a function of α'. In this section we will obtain the general form of the worldvolume action for Dp-branes (in static gauge) from (7.1) in terms of their low-energy field content, i.e. the gauge fields A_μ and scalar fields Φ^m, by using T-duality. Because non-linear electrodynamics is not a very familiar subject to most, let us begin with the following exercise to become acquainted with some of its novel features.

Exercise 7.1. **(a)** *Show that the equations of motion which follow from the Born–Infeld action are given by*

$$\left(\frac{1}{\mathbb{1}-(T^{-1}F)^2}\right)^{\nu\lambda}\partial_\nu F_{\lambda\mu}=0\ .$$

They reduce to the usual Maxwell equations in the field theory limit $\alpha'\to 0$ which decouples all massive string modes.
(b) *Show that in $d=4$ dimensions the Born–Infeld action can be written in the form*

$$S_{\rm BI}^{(d=4)}=\frac{1}{(4\pi^2\alpha')^5\,g_s}\int {\rm d}^4x\,\sqrt{1+\frac{1}{2T^2}\,F_{\mu\nu}F^{\mu\nu}-\frac{1}{16T^4}\left(F_{\mu\nu}\tilde F^{\mu\nu}\right)^2}\ ,$$

where $\tilde F^{\mu\nu}=\frac12\,\epsilon^{\mu\nu\lambda\rho}\,F_{\lambda\rho}$. In this sense the Born–Infeld action interpolates between the Maxwell form $\frac14\,F_{\mu\nu}F^{\mu\nu}$ for small F and the topological density $\frac14\,F_{\mu\nu}\tilde F^{\mu\nu}$ for large F.

(c) *Show that in four spacetime dimensions the Born–Infeld electric field generated by a point charge Q at the origin is given by*

$$E_r=F_{rt}=\frac{Q}{\sqrt{r^4+r_0^4}}\ ,\quad r_0^2=\frac{Q}{T}\ .$$

Thus the distribution $\rho=\frac{1}{4\pi}\,\nabla\cdot\vec E$ of the electric field has an effective radius $r_0\propto\ell_s$.

To transform the Born–Infeld action (7.1) to an action for a Dp-brane with $p<9$, we T-dualize $9-p$ of the ten spacetime directions. Then the $9-p$ directions are described by *Dirichlet* boundary conditions for the open strings, which thereby sit on a $p+1$ dimensional worldvolume hyperplane. We assume that the normal directions x^m, $m=p+1,\dots,9$ are circles which are so small that we can neglect all derivatives along them. The remaining uncompactified worldvolume directions are x^a, $a=0,1,\dots,p$.

Exercise 7.2. *If $\mathcal{M}$ and $\mathcal{N}$ are invertible $p \times p$ and $q \times q$ matrices, respectively, and $\mathcal{A}$ is $p \times q$, show that*

$$\det \begin{pmatrix} \mathcal{N} & -\mathcal{A}^\top \\ \mathcal{A} & \mathcal{M} \end{pmatrix} = \det(\mathcal{M}) \det\left(\mathcal{N} + \mathcal{A}^\top \mathcal{M}^{-1} \mathcal{A}\right) = \det(\mathcal{N}) \det\left(\mathcal{M} + \mathcal{A}\mathcal{N}^{-1} \mathcal{A}^\top\right) .$$

To expand the determinant appearing in (7.1), we apply the determinant formula of Exercise 7.2 with

$$\begin{aligned} \mathcal{N} &= (\eta_{ab} + 2\pi\alpha' F_{ab}) , \\ \mathcal{M} &= (\delta_{mn}) , \\ \mathcal{A} &= (2\pi\alpha' \partial_a A_m) , \end{aligned} \tag{7.2}$$

and use the T-duality rules to replace gauge fields in the T-dual directions by brane coordinates according to

$$2\pi\alpha' A_m = x^m . \tag{7.3}$$

By worldvolume and spacetime reparametrization invariance of the theory, we may choose the "static gauge" in which the worldvolume is aligned with the first $p+1$ spacetime coordinates, leaving $9-p$ transverse coordinates. This amounts to calling the $p+1$ brane coordinates $\xi^a = x^a$, $a = 0, 1, \ldots, p$. In this way the Born–Infeld action can be thereby written as

$$S_{\rm DBI} = -\frac{T_p}{g_s} \int {\rm d}^{p+1}\xi \, \sqrt{- \det_{0 \le a,b \le p} \left(\eta_{ab} + \partial_a x^m \, \partial_b x_m + 2\pi\alpha' F_{ab}\right)} . \tag{7.4}$$

This is known as the "Dirac–Born–Infeld action" and it describes a model of non-linear electrodynamics on a fluctuating p-brane [Leigh (1989)]. The quantity

$$T_p = \frac{1}{\sqrt{\alpha'}} \frac{1}{\left(2\pi\sqrt{\alpha'}\right)^p} \tag{7.5}$$

has dimension mass/volume and is the tension of the p-brane, generalizing the $p=1$ string tension T. The tension formula (7.5) plays a pivotal role in the dynamics of D-branes, as will be discussed in Section 7.3.

To understand the meaning of the action (7.4), let us consider the case where there is no gauge field on the Dp-brane, so that $F_{ab} \equiv 0$. Then the Dirac–Born–Infeld action reduces to

$$S_{\rm DBI}(F=0) = -\frac{T_p}{g_s} \int {\rm d}^{p+1}\xi \, \sqrt{-\det_{a,b} \left(-\eta_{\mu\nu} \, \partial_a x^\mu \, \partial_b x^\nu\right)} \ , \tag{7.6}$$

where we have rewritten the argument of the static gauge determinant in (7.4) in covariant form. The tensor field $h_{ab} = -\eta_{\mu\nu} \, \partial_a x^\mu \, \partial_b x^\nu$ is the induced metric on the worldvolume of the Dp-brane, so that the integrand of (7.6) is the invariant, infinitesimal volume element on the D-brane hypersurface. Thus (7.6) is just the p-brane generalization of the actions we encountered in Chapter 2 for a massive point particle and a string of tension T. So the Dirac–Born–Infeld action is the natural geometric extension, incorporating the worldvolume gauge fields, of the string Nambu–Goto action to the case of D-branes.

7.1.1 *Example*

Example 7.1. To illustrate the utility of describing effects in string theory by using T-duality, and to further give a nice physical origin to the Dirac–Born–Infeld action, let us consider now the example of electric fields in string theory, which have many exotic properties that find their most natural dynamical explanations in the T-dual D-brane picture [Ambjørn *et al* (2003)]. A pure electric background is specified by the field strength tensor

$$F_{0i} = E_i \ , \quad F_{ij} = 0 \ , \tag{7.7}$$

where $i,j = 1,\dots,9$. The Born–Infeld action (7.1) then essentially only involves a simple 2×2 determinant, and it takes the particularly simple form

$$S_{\rm BI}(E) = \frac{1}{g_s} \left(\frac{T}{2\pi}\right)^5 \int {\rm d}^{10}x \, \sqrt{1 - \left(T^{-1}\,\vec{E}\right)^2} \ . \tag{7.8}$$

From (7.8) we see that, at the origin of the source for $\vec{E}$, the electric field attains a *maximum* value

$$E_c = T = \frac{1}{2\pi\alpha'} \ . \tag{7.9}$$

This limiting value arises because for $|\vec{E}| > E_c$ the action (7.8) becomes complex-valued and ceases to make physical sense [Fradkin and Tseytlin

(1985)]. It represents an instability in the system, reflecting the fact that the electromagnetic coupling of open strings is not minimal and creates a divergence due to the fast rising density of string states. Heuristically, since the string effectively carries electric charges of equal sign at each of its endpoints, as $|\vec{E}|$ increases the charges start to repel each other and stretch the string. For field strengths larger than the critical value (7.9), the string tension T can no longer hold the strings together. Note that this instability may be attributed to the Minkowski sign factor of the time direction of the metric $\eta_{\mu\nu}$, and hence it does not arise in a purely magnetic background [Ambjørn *et al* (2003)].

The fact that electric fields in string theory are not completely arbitrary, because they have a limiting value above which the system becomes unstable, is actually *very* natural in the T-dual D-brane picture. For this, let us consider the simplest case of the coupling of an open string to a time-varying but spatially constant electric field $\vec{E} = \partial_0 \vec{A}$. The worldsheet action in the Neumann picture is

$$S_{\rm N} = \frac{1}{4\pi\alpha'} \int {\rm d}^2\xi \ \partial_a x^\mu \, \partial^a x_\mu + {\rm i} \int {\rm d}l \ \vec{A}(x^0) \cdot \partial_\parallel \vec{x} \ , \tag{7.10}$$

where all spacetime coordinate functions x^μ obey Neumann boundary conditions. As we have seen, T-dualizing the nine space directions maps the vector potential $\vec{A}$ onto the trajectory $\vec{y}$ of a D-particle and (7.10) into the Dirichlet picture action

$$S_{\rm D} = \frac{1}{4\pi\alpha'} \int {\rm d}^2\xi \ \partial_a x'^{\,\mu} \, \partial^a x'_\mu + \frac{1}{2\pi\alpha'} \int {\rm d}l \ \vec{y}(x^0) \cdot \partial_\perp \vec{x}\,' \ , \tag{7.11}$$

where $x'^{\,0} = x^0$ still obeys Neumann boundary conditions, while $x'^{\,i}$ obey Dirichlet boundary conditions. The boundary vertex operator in (7.11) creates a moving D0-brane which travels with velocity

$$\vec{v} = \partial_0 \vec{y} = 2\pi\alpha' \, \vec{E} \ . \tag{7.12}$$

In string perturbation theory, the equivalence of the electric field and moving D-brane problems follows from the perturbative duality between Neumann and Dirichlet boundary conditions for the open strings. This is reflected in the equality of the corresponding boundary propagators (see Exercise 6.2)

$$\begin{aligned} \Big\langle \partial_{\sigma_1} x^\mu(\tau_1) \, \partial_{\sigma_2} x^\nu(\tau_2) \Big\rangle_{\rm N} &= - \Big\langle \partial_{\tau_1} x'^{\,\mu}(\tau_1) \, \partial_{\tau_2} x'^{\,\nu}(\tau_2) \Big\rangle_{\rm D} \\ &= \frac{2\alpha' \, \eta^{\mu\nu}}{(\tau_1 - \tau_2)^2} \ . \end{aligned} \tag{7.13}$$

This implies that the open string loop expansions are the same (modulo zero modes). This is true on the boundary of the disc, but *not* in the bulk. The Born–Infeld action (7.8) then simply maps onto the usual action for a relativistic point particle (cf. (2.9)),

$$S_{\mathrm{DBI}}(v) = m \int \mathrm{d}\tau \, \sqrt{1 - \vec{v}^{\,2}} \,, \tag{7.14}$$

where

$$m = T_0 = \frac{1}{g_s \sqrt{\alpha'}} \tag{7.15}$$

is the mass of the D-particle. It follows that, in the dual picture, the existence of a limiting electric field is merely a consequence of the laws of relativistic particle mechanics for a 0-brane, with the "critical" velocity $v_c = 2\pi\alpha' E_c = 1$ corresponding to the speed of light. We note the string coupling dependence of the D0-brane mass (7.15), which reflects the fact that D-branes are really non-perturbative degrees of freedom in superstring theory.

At the velocity v_c, we can make a large Lorentz boost to bring the system to rest, so that in the T-dual picture of the original open string theory the existence of electric fields of strength near (7.9) amounts to a boost to large momentum. Thus string theory with electric background near the critical limit is equivalent to string theory in the infinite momentum frame. This illustrates the overall ease in which things may be interpreted in D-brane language. Put differently, demanding that the D0-branes of Type IIA superstring theory behave as relativistic particles *uniquely* fixes the form of the Born–Infeld action, which is the result of a resummation of all stringy α' corrections. This demonstrates the overall consistency of the Dirac–Born–Infeld action in superstring theory.

7.1.2 *Supergravity Couplings*

Thus far we have been working in *flat* ten dimensional spacetime, which represents a particular background of string theory, i.e. a particular solution to the supergravity equations of motion. It is straightforward, however, to generalize the action (7.4) to *curved* spacetimes, which is tantamount to coupling D-branes to supergravity fields. For this, we incorporate the massless NS–NS spacetime fields of the closed string sector, namely the spacetime metric $g_{\mu\nu}$, the antisymmetric tensor $B_{\mu\nu}$, and the dilaton Φ.

This is done by considering a more general worldsheet action, corresponding to the couplings of the NS–NS fields to the fundamental strings, of the form

$$S = \frac{1}{4\pi\alpha'} \int \mathrm{d}^2\xi \, \left(g_{\mu\nu}\, \partial_a x^\mu \, \partial^a x^\nu + 2\pi\alpha' \, B_{\mu\nu}\, \epsilon^{ab}\, \partial_a x^\mu \, \partial_b x^\nu + \alpha' \, \Phi \, R^{(2)} \right) \,, \tag{7.16}$$

where $R^{(2)}$ is the (scalar) curvature of the two-dimensional worldsheet.

If $B_{\mu\nu}$ is constant, then the familiar B-field coupling in (7.16) is actually a *boundary* term on the disc, since it can then be integrated by parts to give

$$\int \mathrm{d}^2\xi \; B_{\mu\nu}\, \epsilon^{ab}\, \partial_a x^\mu \, \partial_b x^\nu = \int_0^{2\pi} \mathrm{d}\theta \; \frac{1}{2} B_{\mu\nu}\, x^\nu \, \dot{x}^\mu \Big|_{r=1} \,. \tag{7.17}$$

When the electromagnetic coupling in (6.21) with the gauge choice (6.25) is included, the effect of such a B-field term is to shift the field strength $F_{\mu\nu}$ to

$$\mathcal{F}_{\mu\nu} \equiv 2\pi\alpha' \, F_{\mu\nu} - B_{\mu\nu} = 2\pi\alpha' \, (\partial_\mu A_\nu - \partial_\nu A_\mu) - B_{\mu\nu} \,. \tag{7.18}$$

This modification of the B-field is actually required to produce an action which is invariant under the gauge transformations of the antisymmetric tensor field in (3.32), which can be absorbed by shifting the vector potential as

$$A_\mu \longmapsto A_\mu + \frac{1}{2\pi\alpha'} \, \Lambda_\mu \,. \tag{7.19}$$

Thus it is the tensor $\mathcal{F}_{\mu\nu}$ which is the gauge-invariant quantity in the presence of background supergravity fields, and not the gauge field strength $F_{\mu\nu}$.

We can now proceed to perform a derivative expansion of the corresponding disc partition function in exactly the same way we did in Section 6.4. In the slowly-varying field approximation, the modification of the Dirac–Born–Infeld action (7.4) in arbitrary background supergravity fields may then be computed to be [Leigh (1989)]

$$S_{\rm DBI} = -T_p \int \mathrm{d}^{p+1}\xi \; \mathrm{e}^{-\Phi} \sqrt{-\det_{a,b} \left(g_{ab} + B_{ab} + 2\pi\alpha' \, F_{ab} \right)} \,, \tag{7.20}$$

where the string coupling is generated through the relation

$$\frac{1}{g_s} = \mathrm{e}^{-\Phi} \,, \tag{7.21}$$

while g_{ab} and B_{ab} are the pull-backs of the spacetime supergravity fields to the Dp-brane worldvolume. In particular, the induced worldvolume metric is given by

$$\begin{aligned} g_{ab}(\xi) &= g_{\mu\nu}\Big(x(\xi)\Big) \, \frac{\partial x^\mu}{\partial \xi^a} \, \frac{\partial x^\nu}{\partial \xi^b} \\ &= \eta_{ab} + \partial_a x^\mu \, \partial_b x_\mu + \mathcal{O}\Big((\partial x)^4\Big) , \end{aligned} \tag{7.22}$$

where the leading terms in (7.22), which coincide with the induced metric terms in (7.4), come from setting $g_{\mu\nu} = \eta_{\mu\nu}$ in static gauge. As before, F_{ab} is the field strength of the worldvolume $U(1)$ gauge field A_a. The expression (7.20) is now the correct form of the worldvolume action which is spacetime gauge-invariant and also reduces to the appropriate Nambu–Goto type p-brane action (7.6) when $B = F = 0$. It produces an intriguing mixture of gauge theory and gravity on D-branes.

7.2 Supersymmetric Yang–Mills Theory

Let us now expand the *flat* space ($g_{\mu\nu} = \eta_{\mu\nu}$, $B_{\mu\nu} = 0$) action (7.4) for slowly-varying fields to order F^4, $(\partial x)^4$. This is equivalent to passing to the field theory limit $\alpha' \to 0$, which is defined precisely by keeping only degrees of freedom of energy $E \ll \frac{1}{\sqrt{\alpha'}}$, that are observable at length scales $L \gg \ell_s$. In this limit, the infinite tower of massive string states decouples, because such states have masses $m \sim \frac{1}{\sqrt{\alpha'}} \to \infty$ and are thereby energetically unfavourable. Using the formula $\det(\mathcal{A}) = \mathrm{e}^{\,\mathrm{Tr}\,\ln(\mathcal{A})}$, the Dirac–Born–Infeld action can be written as

$$\begin{aligned} S_{\mathrm{DBI}} = &-\frac{T_p\,(2\pi\alpha')^2}{4g_s} \int \mathrm{d}^{p+1}\xi \, \left(F_{ab}F^{ab} + \frac{2}{(2\pi\alpha')^2}\,\partial_a x^m\,\partial^a x_m\right) \\ &- \frac{T_p}{g_s}\,V_{p+1} + \mathcal{O}\left(F^4\right) , \end{aligned} \tag{7.23}$$

where V_{p+1} is the (regulated) p-brane worldvolume. This is the action for a $U(1)$ gauge theory in $p+1$ dimensions with $9-p$ real scalar fields x^m.

But (7.23) is just the action that would result from the dimensional reduction of $U(1)$ Yang–Mills gauge theory (electrodynamics) in ten spacetime dimensions, which is defined by the action

$$S_{\mathrm{YM}} = -\frac{1}{4g_{\mathrm{YM}}^2} \int \mathrm{d}^{10}x \; F_{\mu\nu}F^{\mu\nu} \; . \tag{7.24}$$

Indeed, the ten dimensional gauge theory action (7.24) reduces to the expansion (7.23) of the Dp-brane worldvolume action (up to an irrelevant constant) if we take the fields A_a and $A_m = \frac{1}{2\pi\alpha'}\, x^m$ to depend *only* on the $p+1$ brane coordinates ξ^a, and be independent of the transverse coordinates $x^{p+1}, \dots, x^9$. This requires the identification of the Yang–Mills coupling constant (electric charge) $g_{\rm YM}$ as

$$g_{\rm YM}^2 = g_s\, T_p^{-1}\, (2\pi\alpha')^{-2} = \frac{g_s}{\sqrt{\alpha'}} \left(2\pi \sqrt{\alpha'}\right)^{p-2} . \tag{7.25}$$

For multiple D-branes, while one can derive a non-abelian extension of the Dirac–Born–Infeld action [Myers (1999); Tseytlin (1997)], the technical details would take us beyond the scope of this book. Instead, we will simply make a concise statement about the dynamics of a system of coinciding D-branes, which naturally generalizes the above construction and which also incorporates spacetime supersymmetry [Witten (1996)]. This will be sufficient for our purposes here, and will be treated through the following *axiom*:

The low-energy dynamics of N parallel, coincident Dirichlet p-branes in flat space is described in static gauge by the dimensional reduction to $p+1$ dimensions of $\mathcal{N}=1$ supersymmetric Yang–Mills theory with gauge group $U(N)$ in ten spacetime dimensions.

The ten dimensional action is given by

$$S_{\rm YM} = \frac{1}{4g_{\rm YM}^2} \int {\rm d}^{10}x \left[{\rm Tr}\, \left(F_{\mu\nu}F^{\mu\nu}\right) + 2\,{\rm i}\ {\rm Tr}\, \left(\overline{\psi}\Gamma^\mu\, D_\mu\psi\right)\right] , \tag{7.26}$$

where

$$F_{\mu\nu} = \partial_\mu A_\nu - \partial_\nu A_\mu - {\rm i}\,[A_\mu, A_\nu] \tag{7.27}$$

is the non-abelian field strength of the $U(N)$ gauge field A_μ, and the action of the gauge-covariant derivative D_μ is defined by

$$D_\mu\psi = \partial_\mu\psi - {\rm i}\,[A_\mu, \psi] \ . \tag{7.28}$$

Again, $g_{\rm YM}$ is the Yang–Mills coupling constant, Γ^μ are 16×16 Dirac matrices, and the $N\times N$ Hermitian fermion field ψ is a 16-component Majorana–Weyl spinor of the Lorentz group $SO(1,9)$ which transforms under the adjoint representation of the $U(N)$ gauge group. The field theory (7.26) possesses eight on-shell bosonic, gauge field degrees of freedom, and eight fermionic degrees of freedom after imposition of the Dirac equation $\not{D}\psi = \Gamma^\mu\, D_\mu\psi = 0$.

Exercise 7.3. *Show that the action (7.26) is invariant under the supersymmetry transformations*

$$\delta_\epsilon A_\mu = \frac{\rm i}{2}\,\bar\epsilon\,\Gamma_\mu\psi\ ,$$
$$\delta_\epsilon\psi = -\frac{1}{2}\,F_{\mu\nu}\,[\Gamma^\mu,\Gamma^\nu]\epsilon\ ,$$

where ϵ is an infinitesimal Majorana–Weyl spinor.

Using (7.26) we can construct a supersymmetric Yang–Mills gauge theory in $p+1$ dimensions, with 16 independent supercharges, by dimensional reduction, i.e. we take all fields to be independent of the coordinates $x^{p+1},\dots,x^9$. Then the ten dimensional gauge field A_μ splits into a $p+1$ dimensional $U(N)$ gauge field A_a plus $9-p$ Hermitian scalar fields $\Phi^m = \frac{1}{2\pi\alpha'}\,x^m$ in the adjoint representation of $U(N)$. The Dp-brane action is thereby obtained from the dimensionally reduced field theory as

$$S_{{\rm D}p} = -\frac{T_p\,g_s\,(2\pi\alpha')^2}{4}\int {\rm d}^{p+1}\xi\ {\rm Tr}\,\Big(F_{ab}F^{ab} + 2D_a\Phi^m\,D^a\Phi_m + \sum_{m\neq n}\,[\Phi^m\,,\,\Phi^n]^2 + \text{fermions}\Big)\ , \tag{7.29}$$

where $a,b=0,1,\dots,p$; $m,n=p+1,\dots,9$, and for the moment we do not explicitly display the fermionic contributions. Thus the brane dynamics is described by a supersymmetric Yang–Mills theory on the Dp-brane worldvolume, coupled dynamically to the transverse, adjoint scalar fields Φ^m. This demonstrates, in particular, how to explicitly write the action for the collective coordinates Φ^m representing the fluctuations of the branes transverse to their worldvolume.

Let us consider the Yang–Mills potential in (7.29), which is given by

$$V(\Phi) = \sum_{m \neq n} \mathrm{Tr}\, [\Phi^m \,,\, \Phi^n]^2 \tag{7.30}$$

and is negative definite because $[\Phi^m, \Phi^n]^\dagger = [\Phi^n, \Phi^m] = -[\Phi^m, \Phi^n]$. A classical vacuum of the field theory defined by (7.29) corresponds to a static solution of the equations of motion whereby the potential energy of the system is minimized. It is given by the field configurations which solve simultaneously the equations

$$\begin{aligned} F_{ab} \;=\; D_a\Phi^m \;=\; \psi^\alpha = 0 \,, \\ V(\Phi) = 0 \,. \end{aligned} \tag{7.31}$$

Since (7.30) is a sum of negative terms, its vanishing is equivalent to the conditions

$$[\Phi^m \,,\, \Phi^n] = 0 \tag{7.32}$$

for all m, n and at each point in the $p+1$ dimensional worldvolume of the branes. This implies that the $N \times N$ Hermitian matrix fields Φ^m are simultaneously diagonalizable by a gauge transformation, so that we may write

$$\Phi^m = U \begin{pmatrix} x_1^m & & & 0 \\ & x_2^m & & \\ & & \ddots & \\ 0 & & & x_N^m \end{pmatrix} U^{-1} \,, \tag{7.33}$$

where the $N \times N$ unitary matrix U is *independent* of m. The simultaneous, real eigenvalues x_i^m give the positions of the N distinct D-branes in the m-th transverse direction. It follows that the "moduli space" of classical vacua for the $p+1$ dimensional field theory (7.29) arising from dimensional reduction of supersymmetric Yang–Mills theory in ten dimensions is the quotient space $(\mathbb{R}^{9-p})^N / S_N$, where the factors of $\mathbb{R}$ correspond to the positions of the N D-branes in the $9-p$ dimensional transverse space, and S_N is the symmetric group acting by permutations of the N coordinates x_i. The group S_N corresponds to the residual Weyl symmetry of the $U(N)$ gauge group acting in (7.33), and it represents the permutation symmetry of a system of N *indistinguishable* D-branes.

From (7.29) one can easily deduce that the masses of the fields corresponding to the off-diagonal matrix elements are given precisely by the distances $|x_i - x_j|$ between the corresponding branes. This description means that an interpretation of the D-brane configuration in terms of classical geometry is *only* possible in the classical ground state of the system, whereby the matrices Φ^m are simultaneously diagonalizable and the positions of the individual D-branes may be described through their spectrum of eigenvalues. This gives a simple and natural dynamical mechanism for the appearence of "noncommutative geometry" at short distances [Connes, Douglas and Schwarz (1998); Mavromatos and Szabo (1999); Witten (1996)], where the D-branes cease to have well-defined positions according to classical geometry. Let us now consider an explicit and important example.

7.2.1 *Example*

Example 7.2. The dynamics of N D0-branes in the low-energy limit in flat ten dimensional spacetime is the dimensional reduction of $\mathcal{N}=1$ supersymmetric Yang–Mills theory in ten dimensions to one time direction τ. The ten dimensional gauge field A_μ thereby splits into nine transverse scalars $\Phi^m(\tau)$ and a one dimensional gauge field $A_0(\tau)$ on the worldline. By choosing the gauge $A_0 = 0$, we then get a "supersymmetric matrix quantum mechanics" defined by the action

$$\begin{aligned} S_{\rm D0} = \frac{1}{2g_s\sqrt{\alpha'}} \int {\rm d}\tau\ {\rm Tr}\ \Bigg(\dot{\Phi}^m\,\dot{\Phi}_m + \frac{1}{(2\pi\alpha')^2} \sum_{m<n} [\Phi^m\,,\,\Phi^n]^2 \\ + \frac{1}{2\pi\alpha'}\,\theta^\top\,{\rm i}\,\dot{\theta} - \frac{1}{(2\pi\alpha')^2}\,\theta^\top\,\Gamma_m\,[\Phi^m\,,\,\theta] \Bigg)\,, \end{aligned} \tag{7.34}$$

where Φ^m, $m = 1,\dots,9$ are $N \times N$ Hermitian matrices (with N the number of D0-branes), whose superpartners θ are also $N \times N$ Hermitian matrices which form 16-component spinors under the $SO(9)$ Clifford algebra generated by the 16×16 matrices Γ_m.

The moduli space of classical vacua of the system of N 0-branes is $(\mathbb{R}^9)^N/S_N$, which is simply the configuration space of N identical particles moving in nine dimensional space. But for a general configuration, the matrices only have a geometrical interpretation in terms of a noncommutative geometry. The Yang–Mills D0-brane theory is essentially the nonrelativistic limit of the Born–Infeld D0-brane theory (7.14), obtained by replacing the Lagrangian $m\sqrt{1-\vec{v}^{\,2}}$ with its small velocity limit $-\frac{m\vec{v}^{\,2}}{2}$ for $|\vec{v}| \ll 1$. The $N \to \infty$ limit of this model is believed to describe the non-perturbative dynamics of "M-Theory" and is known as "Matrix Theory" [Banks *et al* (1997); Taylor (2001)].

7.3 Forces Between D-Branes

The final point which we shall address concerning D-brane dynamics is the nature of the interactions between D-branes. For this, we will return to the worldsheet formalism and present the string computation of the static force between two separated Dp-branes. This will also introduce another important string theoretical duality, constituting one of the original dualities that arose in the context of dual resonance models. We will then compare the results of this formalism with the Yang–Mills description that we have developed thus far in this chapter.

We will compute the one-loop open string vacuum amplitude, which is given by the annulus diagram (Fig. 7.1). By using an appropriate modular transformation, this open string graph can be equivalently expressed as the cylinder diagram obtained by "pulling out" the hole of the annulus (Fig. 7.2). In this latter representation, the worldline of the open string boundary is a vertex connecting the vacuum to a single closed string state. We have thereby found that the *one-loop open string* Casimir force is equivalent to a *tree-level closed string* exchange between a pair of D-branes. This equivalence is known as "open-closed string channel duality" and it enables one to straightforwardly identify the appropriate interaction amplitudes for D-branes [Polchinski (1995)]. In particular, two Dp-branes interact gravitationally by exchanging closed strings corresponding, in the massless sector, to graviton and dilaton states.

As is usual in quantum field theory, the one-loop vacuum amplitude $\mathcal{A}$ is given by the logarithm of the partition function $Z_{\rm vac}$ determined as the fluctuation determinant of the full theory arising from integrating out the free worldsheet fields. With L_0 the worldsheet Hamiltonian, we thereby have

$$\begin{aligned}\mathcal{A} = \ln(Z_{\rm vac}) &= \ln\left(P_{\rm GSO}\,\frac{1}{\sqrt{\det_{{\rm NS}\oplus{\rm R}}(L_0-a)}}\right)\\ &= -\frac{1}{2}\,{\rm Tr}_{\,{\rm NS}\oplus{\rm R}}\Big[P_{\rm GSO}\,\ln(L_0-a)\Big]\\ &= -\frac{V_{p+1}}{2}\int\frac{{\rm d}^{p+1}k}{(2\pi)^{p+1}}\,{\rm tr}_{\,{\rm NS}\oplus{\rm R}}\Big[P_{\rm GSO}\,\ln\left(k^2+m^2\right)\Big]\ ,\end{aligned} \tag{7.35}$$

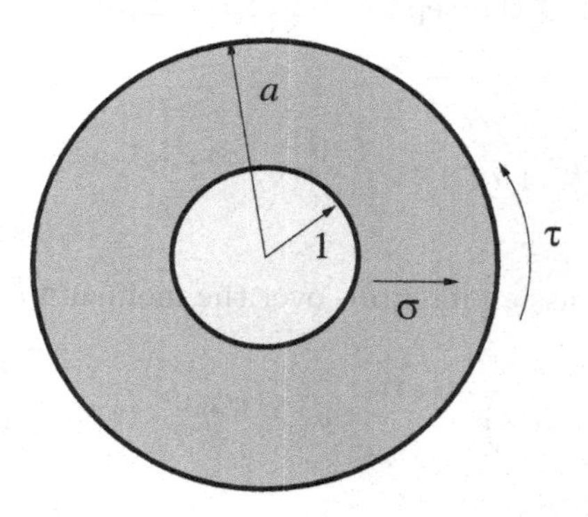

Fig. 7.1 The annulus diagram, whose local coordinates are $(\xi^0, \xi^1) = (\sigma, \tau)$ with $0 \leq \sigma \leq \pi$, $0 \leq \tau \leq 2\pi t$ and modulus $0 \leq t < \infty$. The inner radius of the annulus is set to unity by a conformal transformation, while the outer radius is $a = \mathrm{e}^{-t}$. Here $\xi^1 = \tau$ is the worldsheet time coordinate, so that the boundaries represent a pair of D-branes with an open string stretched between them.

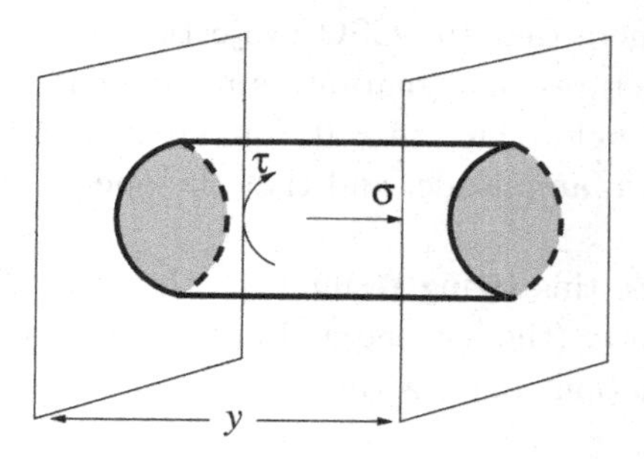

Fig. 7.2 The cylinder diagram as a modular transform of the annulus graph. The separation between the two D-branes is y. Now $\xi^0 = \sigma$ is the worldsheet time coordinate and the diagram represents a single closed string propagating in the tree channel from one D-brane to the other.

where k are the zero-mode momenta in the Neumann directions (which vanish along the Dirichlet directions), the worldvolume factor is inserted to make the momentum integrations dimensionless, and the open string mass spectrum is given by (cf. (6.13))

$$m^2 = \frac{1}{\alpha'}\Big(N - a\Big) + \left(\frac{y}{2\pi\alpha'}\right)^2 \tag{7.36}$$

with y^n, $n = p+1, \ldots, 9$ the separation of the Dp-branes. By using the elementary identity

$$-\frac{1}{2}\ln\left(k^2+m^2\right) = \int_0^\infty \frac{\mathrm{d}t}{2t}\ \mathrm{e}^{-2\pi\alpha'\,(k^2+m^2)\,t}\ , \tag{7.37}$$

we may perform the Gaussian integrals over the momenta k in (7.35) to get

$$\mathcal{A} = -2V_{p+1}\int_0^\infty \frac{\mathrm{d}t}{2t}\left(8\pi^2\alpha' t\right)^{-(p+1)/2}\ \mathrm{e}^{-y^2\,t/2\pi\alpha'}\ \mathrm{tr}_{\,\mathrm{NS}\oplus\mathrm{R}}\left(P_{\mathrm{GSO}}\,q^{N-a}\right)\ , \tag{7.38}$$

where

$$q \equiv \mathrm{e}^{-2\pi t}\ . \tag{7.39}$$

The trace that appears in (7.38) is exactly the *same* sort of trace that we encounterd in the calculation of the one-loop closed superstring amplitude in Section 4.5, with the identification $t = -\mathrm{i}\,\tau$ (see (4.53), p. 56). In particular, we found there that the GSO projection, giving the appropriate sum over spin structures that guarantees modular invariance, yields a *vanishing* result. We therefore have $\mathcal{A} = 0$, consistent with the fact that we are computing a vacuum amplitude, and that the open string spectrum is supersymmetric.

Let us now compare this string result with the static force computed from quantum field theory (the low-energy limit). Among many others, the Type II supergravity action contains the terms

$$\begin{aligned} S_{\mathrm{RR}} = &-\frac{1}{2\kappa^2}\int \mathrm{d}^{10}x\ F^{(p+2)}_{\mu_1\cdots\mu_{p+2}}F^{(p+2)\,\mu_1\cdots\mu_{p+2}} \\ &+ q_p\int \mathrm{d}^{p+1}\xi\ C^{(p+1)}_{a_1\cdots a_{p+1}}\ \epsilon^{a_1\cdots a_{p+1}}\ , \end{aligned} \tag{7.40}$$

where κ is the gravitational constant, q_p is the charge of the Dp-brane under the Ramond–Ramond $p+1$-form potential $C^{(p+1)}$, and

$$C^{(p+1)}_{a_1\cdots a_{p+1}} = C^{(p+1)}_{\mu_1\cdots\mu_{p+1}}\ \frac{\partial x^{\mu_1}}{\partial \xi^{a_1}}\cdots\frac{\partial x^{\mu_{p+1}}}{\partial \xi^{a_{p+1}}} \tag{7.41}$$

is the "pull-back" of $C^{(p+1)}$ to the Dp-brane worldvolume. A very tedious (but standard) perturbative calculation in the $p+1$-form quantum field theory defined by the action (7.40) establishes that the corresponding vacuum energy vanishes *provided* we make the identification [Polchinski (1995)]

$$\boxed{T_p = q_p \ .} \tag{7.42}$$

This coincidence between Dp-brane tension and Ramond–Ramond charge is one of the most important results in D-brane physics, and indeed it was one of the sparks which ignited the second superstring revolution. It implies that the Ramond–Ramond repulsion between identical, parallel Dp-branes cancels exactly their gravitational and dilaton attraction. The cancellation of the static force is a consequence of spacetime supersymmetry, and it forces us to accept the fact that D-branes *are* the Type II R–R charged states [Polchinski (1995)].

7.3.1 *BPS States*

The remarkable properties of D-brane interactions that we have discovered above arise because D-branes describe certain special, non-perturbative extended states of the Type II superstring which carry non-trivial R–R charge. To understand this point better, we recall that Type II superstring theory in the bulk (away from D-branes) possesses $\mathcal{N} = 2$ spacetime supersymmetry. However, the open string boundary conditions are invariant under only one of these supersymmetries. In other words, the Type II vacuum without D-branes is invariant under $\mathcal{N} = 2$ supersymmetry, but the state containing a D-brane is invariant only under $\mathcal{N} = 1$ [Polchinski (1995)]. So a D-brane may be characterized as a state which preserves only *half* of the original spacetime supersymmetries. Such a state is known as a "Bogomol'ny–Prasad–Sommerfeld (BPS) state" [Figueroa-O'Farrill (2001)].

Generally, BPS states carry conserved charges which are determined entirely by their mass in the corresponding (extended) supersymmetry algebra. In the present case, there is only one set of charges with the correct Lorentz transformation properties, namely the antisymmetric Ramond–Ramond charges. So unlike the fundamental string, a D-brane carries R–R charge, *consistent* with the fact that it is a BPS state. This property is known explicitly from the realizations of D-branes as solitonic solutions of the classical supergravity equations of motion [Duff, Khuri and Lu (1995)].

Such BPS bound state configurations of D-branes can also be realized in their low-energy, supersymmetric Yang–Mills theory description. The corresponding BPS energies of these systems agree with the supersymmetric Yang–Mills energy given by

$$E_{\rm YM} = \frac{\pi^2\,\alpha'\,T_p}{g_s}\int {\rm d}^{p+1}\xi\ \ {\rm Tr}\left(F_{ab}F^{ab}\right)\ . \tag{7.43}$$

The BPS property is protected at the quantum level by supersymmetry, via the usual non-renormalization theorems. Thus the relation (7.42) between the mass and charge of a D-brane cannot be modified by any perturbative or non-perturbative effects. The fact that D-branes fill out supermultiplets of the underlying supersymmetry algebra has been the crucial property in testing the various duality conjectures that were discussed in Chapter 1.

Chapter 8

Ramond–Ramond Couplings of D-Branes

The final ingredient we need in our description of the dynamics of D-branes is a determination of how the worldvolume fields on a D-brane couple to supergravity fields of the Ramond–Ramond sector. Recall that a Dp-brane carries R–R charge, which means that it couples to the R–R $p+1$-form potential $C^{(p+1)}$ through the integral $\int_{\Sigma_{p+1}} C^{(p+1)}$, where Σ_{p+1} denotes the $p+1$ dimensional worldvolume swept out as the D-brane propagates through spacetime. It turns out that this topological coupling is only a small piece of a more complicated topologically invariant action which couples the R–R potentials to the gauge (and transverse scalar) fields, and also to gravity on the brane. We will begin by showing that there is an *anomaly* arising in the generically chiral worldvolume field theory on the *intersection* of two or more D-branes. Then the complete coupling of a Dp-brane can be derived by demanding that it cancels the possible anomalies on the branes through the standard inflow mechanism. Indeed, as we shall find, all of these induced couplings are anomalous with respect to gauge transformations of the background, and they lead to somewhat striking properties of the branes themselves.

8.1 D-Brane Anomalies

It is straightforward to determine what kinds of anomalous fields can appear on brane-brane intersections [Green, Harvey and Moore (1997)]. It turns out that the *only* massless anomalous field that can arise is a charged chiral spinor field in the Ramond sector, reduced from $d = 10$ spacetime dimensions to $p+1$ dimensions. We will work in Type IIB superstring theory, so that $\dim(\Sigma_{p+1}) = p+1$ is even. Via the Green–Schwarz mechanism, the $d = 10$ Type IIB theory is chiral but anomaly free.

The corresponding result for Type IIA D-branes can then be deduced by applying T-duality.

Let us compute the anomaly for a chiral spinor field reduced from the ten dimensional spacetime manifold X to the Dp-brane worldvolume submanifold $\Sigma = \Sigma_{p+1} \subset X$. Under the dimensional reduction $X \to \Sigma$, the local Lorentz group of X is broken to the worldvolume Lorentz group plus an R-symmetry group corresponding to rotations in the transverse space,

$$SO(9,1) \longrightarrow SO(p,1) \times SO(9-p) . \tag{8.1}$$

The first factor is the structure group of the tangent bundle $T\Sigma$ over the brane worldvolume Σ, while the second factor corresponds to the structure group of the normal bundle $N\Sigma$ to Σ in X. The spacetime tangent bundle thereby decomposes near Σ as the Whitney sum

$$TX\big|_{\Sigma} = T\Sigma \oplus N\Sigma . \tag{8.2}$$

Correspondingly, a field in ten dimensions in some representation R of $SO(9,1)$ will decompose into various multiplets of fields in $p+1$ dimensions, and in representations (R_t^i, R_n^i) of $SO(p,1) \times SO(9-p)$. From a mathematical perspective, this means that a section of TX in some representation R will decompose into sections of $T\Sigma \otimes N\Sigma$ in representations $R_t^i \otimes R_n^i$. The first factors correspond to gravity on the brane viewed as a gauge theory whose gauge field is the spin connection on $T\Sigma$, transforming under the R_t^i's. The second factor corresponds to a gauge symmetry whose gauge field is the spin connection on $N\Sigma$, transforming under the R_n^i's.

The anomaly on the D-brane is encoded in a closed, gauge invariant inhomogeneous differential form I which is a function of the curvature two-forms of the gauge, tangent and normal bundles over the brane. Globally, it is of the form

$$I = I_0 + \mathrm{d}I^{(0)} , \tag{8.3}$$

where $I^{(0)}$ is *not* gauge invariant. Its gauge variation is related to the first Wess–Zumino descendent $I^{(1)}$ by

$$\delta_g I^{(0)} = \mathrm{d}I^{(1)} . \tag{8.4}$$

As usual in quantum field theory [Alvarez-Gaumé and Ginsparg (1985)], the anomaly then takes the form

$$\mathcal{A} = 2\pi\,\mathrm{i} \int_{\Sigma} I^{(1)} , \tag{8.5}$$

which ensures that the Wess–Zumino consistency condition is automatically satisfied.

It is well-known [Alvarez-Gaumé and Ginsparg (1985)] that the index of the Dirac operator on X gives the perturbative chiral gauge anomaly of a Dirac spinor field on X. We can thereby use the standard Atiyah–Singer index theorem to compute the anomaly for a chiral spinor field propagating on $\Sigma \subset X$. Given the Whitney sum decomposition (8.2), the corresponding positive and negative chirality spin bundles $S^{\pm}_{TX}$ decompose in the usual way in terms of the tensor product $\mathbb{Z}_2$-grading of the spin bundles $S^{\pm}_{T\Sigma}$ and $S^{\pm}_{N\Sigma}$ as

$$S_{TX} = S^{+}_{TX} \oplus S^{-}_{TX} \quad \longrightarrow \quad S_{T\Sigma} \otimes S_{N\Sigma} = \left(S^{+}_{T\Sigma} \oplus S^{-}_{T\Sigma}\right) \otimes \left(S^{+}_{N\Sigma} \oplus S^{-}_{N\Sigma}\right) \ . \tag{8.6}$$

In other words, under the dimensional reduction $X \to \Sigma$ the chiral spinor bundles over spacetime decompose as

$$S^{\pm}_{TX} \quad \longrightarrow \quad \left(S^{\pm}_{T\Sigma} \otimes S^{+}_{N\Sigma}\right) \oplus \left(S^{\mp}_{T\Sigma} \otimes S^{-}_{N\Sigma}\right) \ . \tag{8.7}$$

The Dirac operator $\mathrm{i}\,\not{D}$ for the charged, reduced fermion field acts on sections of the bundle

$$E = \left(S_{T\Sigma} \otimes S_{N\Sigma}\right) \otimes V_{\rho} = E^{+} \oplus E^{-} \ , \tag{8.8}$$

where V_{ρ} is the Chan–Paton gauge bundle $V \to \Sigma$ in the representation ρ under which the fermion field transforms and

$$E^{\pm} = \Big(\left(S^{\pm}_{T\Sigma} \otimes S^{+}_{N\Sigma}\right) \oplus \left(S^{\mp}_{T\Sigma} \otimes S^{-}_{N\Sigma}\right)\Big) \otimes V_{\rho} \ . \tag{8.9}$$

More precisely, $\mathrm{i}\,\not{D}$ defines the (trivial) two-term complex

$$\mathrm{i}\,\not{D} \ : \ \Gamma^{\infty}(\Sigma, E^{+}) \ \longrightarrow \ \Gamma^{\infty}(\Sigma, E^{-}) \ . \tag{8.10}$$

The index $\mathrm{ind}(\mathrm{i}\,\not{D})$ of the chiral Dirac operator (8.10) is defined to be the difference between the dimensions of its kernel and cokernel,

$$\mathrm{ind}(\mathrm{i}\,\not{D}) = \dim\ker(\mathrm{i}\,\not{D}) - \dim\ker(\mathrm{i}\,\not{D}^{\dagger}) \ . \tag{8.11}$$

Since non-zero positive and negative chirality eigenstates of $\mathrm{i}\,\not{D}$ are paired by the appropriate chirality operator, the integer (8.11) computes the difference between the numbers of zero energy eigenstates of positive and negative chirality, i.e. the spectral asymmetry of the corresponding Hamiltonian representing a potential breakdown of the chiral symmetry. The usual index theorem applied to the two-term complex (8.10) then yields [Alvarez-Gaumé and Ginsparg (1985); Scrucca and Serone (1999)]

$$\mathrm{ind}(\mathrm{i}\,\not{D}) = (-1)^{\frac{(p+1)(p+2)}{2}} \int_\Sigma \mathrm{ch}(E) \wedge \frac{\mathrm{Td}(T_{\mathbb{C}}\Sigma)}{e(T\Sigma)} \ , \tag{8.12}$$

where ch denotes the Chern character, e is the Euler class, Td is the Todd class, and $T_{\mathbb{C}}\Sigma = T\Sigma \otimes \mathbb{C}$ is the complexified tangent bundle of the world-volume Σ. We assume throughout that Σ is a spin manifold, but the extension to the spin^c case is straightforward. Indeed, it is known that D-brane worldvolume manifolds always admit at least spin^c structures [Bryant and Sharpe (1999)].

For the present purposes, we can rewrite the index formula (8.12) using various properties of the characteristic classes. First of all, we can relate the Todd class to the Dirac $\widehat{A}$-genus through

$$\widehat{A}(T\Sigma) = \sqrt{\mathrm{Td}(T_{\mathbb{C}}\Sigma)} \tag{8.13}$$

by using the spin structure on Σ. Next, we apply the additivity and multiplicativity properties of the Chern character to write

$$\begin{aligned} \mathrm{ch}(V \oplus W) &= \mathrm{ch}(V) + \mathrm{ch}(W) \ , \\ \mathrm{ch}(V \otimes W) &= \mathrm{ch}(V) \wedge \mathrm{ch}(W) \ . \end{aligned} \tag{8.14}$$

If $V \to \Sigma$ is any oriented real spin bundle, then the Chern character of its spin cover is given by

$$\mathrm{ch}(S_V) = \frac{e(V)}{\widehat{A}(V)} \ . \tag{8.15}$$

Finally, for unitary gauge bundles $V \to \Sigma$, we have

$$\begin{aligned} \mathrm{ch}_{\rho_1 \otimes \rho_2}(V) &= \mathrm{ch}_{\rho_1}(V) \wedge \mathrm{ch}_{\rho_2}(V) \ , \\ \mathrm{ch}_{\overline{\rho}}(V) &= \mathrm{ch}(\overline{V}) \end{aligned} \tag{8.16}$$

where $\mathrm{ch}_\rho(V) = \mathrm{ch}(V_\rho)$. We are interested in the usual Chan–Paton representation $\rho = \mathbf{N} \otimes \overline{\mathbf{N}}$ of the open string gauge group $U(N)$. In this way, we may rewrite (8.12) after a bit of algebra as

$$\boxed{\mathrm{ind}(\mathrm{i}\,\not{D}) = \int_\Sigma \mathrm{ch}(V) \wedge \mathrm{ch}(\overline{V}) \wedge \frac{\widehat{A}(T\Sigma)}{\widehat{A}(N\Sigma)} \wedge e(N\Sigma) \ .} \tag{8.17}$$

The integration in (8.17) can be written explicitly in terms of differential forms as follows. If A is any gauge field connection of the $U(N)$ Chan–Paton gauge bundle $V \to \Sigma$, then the Chern characteristic class may be represented in terms of the closed (inhomogeneous) differential form

$$\mathrm{ch}(V) = \mathrm{ch}(F) = \mathrm{Tr}\, \exp\left(\frac{F}{2\pi}\right) , \tag{8.18}$$

where Tr is the trace in the fundamental representation $\mathbf{N}$ of $U(N)$, and $F = \mathrm{d}A + A \wedge A$ is the curvature field strength of the gauge field A. On the other hand, the $\widehat{A}$-genus characteristic classes can be represented in terms of the Riemann curvature two-form $(R_V)_{ab}$ of the given bundle $V \to \Sigma$ as

$$\widehat{A}(V) = \widehat{A}(R_V) = \prod_{l \geq 1} \frac{\lambda_l}{\sinh \lambda_l} , \tag{8.19}$$

where $4\pi\lambda_l$ are the skew-eigenvalues of $(R_V)_{ab}$.

The index formula (8.17) gives the anomaly action

$$\mathcal{A} = 2\pi\,\mathrm{i}\ \mathrm{ind}(\mathrm{i}\,\not{D})^{(1)} . \tag{8.20}$$

In the next section we will match this anomaly term with that which arises in a given brane-brane intersection. By matching the index theoretical result (8.17), (8.20) with the general form (8.5) of the anomaly, we see that the invariant anomaly polynomials for the spinor fields on overlapping D-branes is given by

$$\boxed{I(F, R_T, R_N) = \mathrm{ch}(F) \wedge \mathrm{ch}(-F) \wedge \frac{\widehat{A}(R_T)}{\widehat{A}(R_N)} \wedge e(R_N) .} \tag{8.21}$$

8.2 Chern–Simons Actions

We will now show that the D-brane anomaly computed in the previous section is cancelled exactly by the inflow of anomaly associated to certain (anomalous) couplings of the R–R tensor fields. For this, we consider a collection of D-branes Σ_i in the spacetime X and postulate Ramond–Ramond couplings of the form

$$S_{\mathrm{int}} = \sum_i \frac{\mu_i}{2} \int_{\Sigma_i} C \wedge Y_i , \tag{8.22}$$

where $C = \sum_p C^{(p+1)}$ is the total inhomogeneous R–R potential, and Y_i (to be determined below) will be an invariant function of the gauge and gravitational curvatures on Σ_i. Via an integration by parts, the integrand in (8.22) can be written in terms of the constant parts $Y_{i0} = 1$ (by suitably normalizing the charge coupling constants μ_i) and the descendents $Y_i^{(0)}$ as $C - G \wedge Y_i^{(0)}$, where $G = \mathrm{d}C$ is the total R–R field strength. Recall that we are working in the Type IIB case, so that C is of even $\mathbb{Z}_2$ form degree.

The complete action for the Ramond–Ramond fields in the presence of sources (i.e. D-branes) in then a modification of that written in (7.40) and can be expressed as an integral over X as

$$S_{\mathrm{RR}} = -\frac{1}{4} \int_X G \wedge *G - \sum_i \frac{\mu_i}{2} \int_X \delta_{\Sigma_i} \wedge \left(C - G \wedge Y_i^{(0)}\right) , \tag{8.23}$$

where the first term is the usual supergravity kinetic term for the form fields $G = \mathrm{d}C$, $*$ denotes the Hodge duality operator on X, and δ_{Σ_i} is the de Rham current representative of the Poincaré dual in X to the forms on Σ_i. The rank of δ_{Σ_i} is the codimension $9 - p_i$ of Σ_i in X, and locally it is delta-function supported in the transverse space to Σ_i as

$$\delta_{\Sigma_i} = \delta(x^{p_i+1}) \cdots \delta(x^9) \, \mathrm{d}x^{p_i+1} \wedge \cdots \wedge \mathrm{d}x^9 . \tag{8.24}$$

However, *globally* it is a section of the normal bundle $N\Sigma_i$ to Σ_i in X.

The action (8.23) implies the equations of motion and Bianchi identity

$$\begin{aligned} \mathrm{d} * G &= \sum_i \mu_i \, \delta_{\Sigma_i} \wedge Y_i , \\ \mathrm{d}G &= -\sum_i \mu_i \, \delta_{\Sigma_i} \wedge \overline{Y_i} , \end{aligned} \tag{8.25}$$

where $\overline{Y_i}$ is obtained from Y_i by complex conjugation of the open string gauge group representation $\rho = \mathbf{N} \otimes \overline{\mathbf{N}}$. Recall from Section 5.1.2 that the relation $G = \mathrm{d}C$ between R–R fields and potentials was derived from the global condition $\mathrm{d} * G = \mathrm{d}G = 0$ in the absence of D-branes. From (8.25) it follows that their presence alters this requirement globally, so that $G \neq \mathrm{d}C$. The minimal solution of the Bianchi identity is

$$G = \mathrm{d}C - \sum_i \mu_i \, \delta_{\Sigma_i} \wedge \overline{Y_i}^{(0)} . \tag{8.26}$$

The Ramond–Ramond field G is the bonafide physical observable, so we will demand that it be gauge invariant,

$$\delta_g G = 0 , \tag{8.27}$$

just like one demands that Yang–Mills field strengths be essentially gauge invariant. From (8.26) it then follows that the Ramond–Ramond potential C acquires an anomalous gauge transformation, in order to compensate the gauge variation of the second term, of the form

$$\delta_g C = \sum_i \mu_i \, \delta_{\Sigma_i} \wedge \overline{Y_i}^{(1)} . \tag{8.28}$$

After another integration by parts, it follows that the gauge variation of the supergravity action (8.23) is given by

$$\delta_g S_{\rm RR} = -\sum_{i,j} \frac{\mu_i \mu_j}{2} \int\limits_X \delta_{\Sigma_i} \wedge \delta_{\Sigma_j} \wedge \big(Y_i \wedge \overline{Y_j} \big)^{(1)} . \tag{8.29}$$

It is a standard mathematical result [Cheung and Yin (1998); Scrucca and Serone (1999)] that in cohomology δ_{Σ_i} may be identified with the Thom class $\Phi[N\Sigma_i]$ of the normal bundle $N\Sigma_i$, whose zero section is the Euler class of $N\Sigma_i$. It follows that the product of de Rham currents in (8.29) can be written as

$$\delta_{\Sigma_i} \wedge \delta_{\Sigma_j} = \delta_{\Sigma_{ij}} \wedge e(N\Sigma_{ij}) \tag{8.30}$$

where $\Sigma_{ij} = \Sigma_i \cap \Sigma_j$. Using the degree of freedom left over in the descent procedure, we may thereby write the gauge variation (8.29) as

$$\delta_g S_{\rm RR} = -\sum_{i,j} \frac{\mu_i \mu_j}{2} \int\limits_{\Sigma_{ij}} \Big(Y_i \wedge \overline{Y_j} \wedge e(N\Sigma_{ij}) \Big)^{(1)} . \tag{8.31}$$

From (8.31) we see that all of the gauge anomaly is localized on the D-brane intersections.

The anomaly inflow can therefore be written as

$$\mathcal{A}_{ij} = 2\pi\, {\rm i} \int\limits_{\Sigma_{ij}} I_{ij}^{(1)} , \tag{8.32}$$

where

$$I_{ij} = -\frac{\mu_i \mu_j}{4\pi} \, Y_i \wedge \overline{Y_j} \wedge e(N\Sigma_{ij}) . \tag{8.33}$$

By matching this with the index theory result (8.21), we see that the two anomalies cancel each other if

$$Y_i(F,R_T,R_N) = \frac{\sqrt{4\pi}}{\mu_i}\,\mathrm{ch}(F)\wedge\sqrt{\frac{\widehat{A}(R_T)}{\widehat{A}(R_N)}}\ . \tag{8.34}$$

By adjusting the charges μ_i in (8.22) to correctly account for the tension of a D-brane, we then find that the complete coupling of a Dp-brane to a Ramond–Ramond field is given by the action

$$\boxed{S_{\rm CS} = \frac{T_p}{2}\int\limits_{\Sigma_{p+1}} C\wedge\mathrm{ch}(F)\wedge\sqrt{\frac{\widehat{A}(R_T)}{\widehat{A}(R_N)}}\ .} \tag{8.35}$$

This is the celebrated "Chern–Simons action" on a D-brane [Douglas (1999)].

8.3 Branes within Branes

The Chern–Simons action (8.35) is a fundamental ingredient in the description of (low-energy) D-brane dynamics, and it is at the heart of various unexpected constructions involving branes of differing dimensions [Douglas (1999)]. To illustrate these novel phenomena, let us for simplicity consider only topologically trivial brane worldvolume manifolds and set $\widehat{A}(R_T) = \widehat{A}(R_N) = 1$. Then the Ramond–Ramond couplings (8.35) are given by

$$S_{\rm CS} = \frac{T_p}{2}\int\limits_{\Sigma_{p+1}} C\wedge\mathrm{Tr}\left(\mathrm{e}^{F/2\pi}\right)\ . \tag{8.36}$$

The total R–R potential is $C=\sum_p C^{(p+1)}$ with p odd for Type IIB and p even for Type IIA superstrings. We expand the exponential of the curvature two-form F in (8.36), keeping at each order only those powers which pair with the appropriate form degree in C to yield a non-vanishing top-form (of degree $p+1$) when integrated over the $p+1$ dimensional worldvolume

Σ_{p+1}. This gives

$$\begin{aligned} S_{\rm CS} = \frac{N\,T_p}{2} \int\limits_{\Sigma_{p+1}} C^{(p+1)} + \frac{T_p}{4\pi} \int\limits_{\Sigma_{p+1}} C^{(p-1)} \wedge {\rm Tr}\,(F) \\ + \frac{T_p}{8\pi^2} \int\limits_{\Sigma_{p+1}} C^{(p-3)} \wedge {\rm Tr}\,(F \wedge F) \\ + \cdots + \frac{T_p}{2k!\,(2\pi)^k} \int\limits_{\Sigma_{p+1}} C^{(p+1-2k)} \wedge {\rm Tr}\,(\underbrace{F \wedge F \wedge \cdots \wedge F}_{k \text{ times}}) + \cdots . \end{aligned} \tag{8.37}$$

Each term in the series (8.37) has a profound physical interpretation. The first term represents the usual p-brane charge as in (7.40), with the correct factor ${\rm Tr}\ \mathbb{1} = N$ appropriate to N coincident Dp-branes. The second term, coupling to the Ramond–Ramond potential $C^{(p-1)}$, carries $p-2$-brane charge $\frac{1}{2\pi}\int_{S^2_\infty} {\rm Tr}\,(F)$, where S^2_∞ is a two-dimensional sphere at infinity inside the original Dp-brane worldvolume Σ_{p+1}. It is induced by turning on a non-trivial magnetic flux (which is necessarily quantized on a two-sphere) for the gauge fields carried by the brane. In a similar vein, the third term corresponds to $p-4$-brane charge $\frac{1}{8\pi^2}\int_{S^4_\infty} {\rm Tr}\,(F \wedge F)$ and is induced by turning on a non-trivial instanton charge for the worldvolume gauge fields. The k-th term in the series represents $p-2k$-brane charge $\frac{1}{k!\,(2\pi)^k}\int_{S^{2k}_\infty} {\rm Tr}\,(F \wedge^k F)$ realized as a topological Pontryagin charge in degree k for the Chan–Paton gauge fields.

Generally, the integral cohomology class proportional to $F \wedge^k F$ is the Poincaré dual of a $p-2k$ dimensional homology class which describes a system of embedded $p-2k$-branes in the Dp-branes with worldvolume Σ_{p+1}. In other words, the integral form corresponding to $F \wedge^k F$ carries $p-2k$-brane charge, because it couples to the appropriate R–R potential $C^{(p-2k+1)}$ which we used in Section 5.1.3 to define the given Ramond–Ramond charges. Thus when the Chan–Paton gauge bundle over the worldvolume of N coincident Dirichlet p-branes is topologically non-trivial, the gauge field configuration carries R–R charges associated with D-branes of dimension $< p$. This gives us realizations of D-branes as conventional gauge theoretic solitons in higher dimensional branes. For example, one can realize a D$(p-2)$-brane inside a Dp-brane as a magnetic vortex or monopole, a D$(p-4)$-brane inside a Dp-brane as an instanton, and so on. An analogous story comes about from considering non-trivial Riemann curvature couplings associated to gravitation on the D-branes [Cheung and Yin (1998); Scrucca and Serone (1999)].

Chapter 9

Solutions to Exercises

Chapter 2

2.1.

$$
\begin{aligned}
0 &= \frac{\delta S[x]}{\delta x^\nu} \\
&= \int \mathrm{d}\tau \ -\frac{m}{2}\left(-\dot{x}^\mu \dot{x}_\mu\right)^{-1/2} \frac{\delta}{\delta x^\nu}\left(-\dot{x}^\mu \dot{x}_\mu\right) \\
&= \int \mathrm{d}\tau \ m\left(\dot{x}^\mu \dot{x}_\mu\right)^{-1/2} \dot{x}_\mu \frac{\mathrm{d}}{\mathrm{d}\tau} \delta^\mu{}_\nu \\
&= \int \mathrm{d}\tau \ -\frac{\mathrm{d}}{\mathrm{d}\tau}\left(\frac{m\,\dot{x}_\nu}{\sqrt{\dot{x}^\mu \dot{x}_\mu}}\right) + \dots
\end{aligned}
$$

where the ellipsis denotes total τ-derivative terms.

2.2.

Under $\xi \mapsto \xi(\xi')$, one has

$$
\partial_a = \frac{\partial}{\partial \xi^a} = \frac{\partial \xi'^{\,b}}{\partial \xi^a} \frac{\partial}{\partial \xi'^{\,b}} \equiv \frac{\partial \xi'^{\,b}}{\partial \xi^a}\, \partial'_b
$$

by the chain rule. Using $\det(A\,B) = (\det A)\,(\det B)$ we get

$$
\det_{a,b}\left(\partial_a x^\mu\, \partial_b x_\mu\right) = \left|\frac{\partial \xi'}{\partial \xi}\right|^2 \det_{a,b}\left(\partial'_a x^\mu\, \partial'_b x_\mu\right)
$$

where $\left|\frac{\partial \xi'}{\partial \xi}\right|$ is the Jacobian of the reparametrization $\xi \mapsto \xi(\xi')$. The substitution formula

$$\mathrm{d}^2\xi = \mathrm{d}^2\xi' \left|\frac{\partial \xi}{\partial \xi'}\right| = \mathrm{d}^2\xi' \left|\frac{\partial \xi'}{\partial \xi}\right|^{-1}$$

then gives the required result.

2.3.

Using

$$\det_{a,b}\left(\gamma^{ab}\right) = \exp\left(\operatorname{tr}_{a,b}\left(\log \gamma^{ab}\right)\right)$$

we have

$$\begin{aligned}\frac{\delta}{\delta\gamma^{ab}} \det_{a,b}\left(\gamma_{ab}\right) &= \frac{\delta}{\delta\gamma^{ab}} \exp\left(-\operatorname{tr}_{a,b}\left(\log \gamma^{ab}\right)\right) \\ &= -\gamma_{ab} \exp\left(-\operatorname{tr}_{a,b}\left(\log \gamma^{ab}\right)\right) \\ &= -\gamma_{ab}\, \gamma\ ,\end{aligned}$$

and so

$$\begin{aligned} 0 &= \frac{\delta S[x,\gamma]}{\delta\gamma^{ab}} \\ &= \int \mathrm{d}^2\xi\ -T\left(\sqrt{-\gamma}\, h_{ab} - \frac{1}{2}\frac{\delta}{\delta\gamma^{ab}}(\gamma)\,\frac{1}{\sqrt{-\gamma}}\,\gamma^{cd}\, h_{cd}\right) \\ &= \int \mathrm{d}^2\xi\ -T\left(\sqrt{-\gamma}\, h_{ab} - \frac{1}{2}\gamma_{ab}\,(-\gamma)\,\frac{\gamma^{cd}\, h_{cd}}{\sqrt{-\gamma}}\right) \\ &= \int \mathrm{d}^2\xi\ -T\sqrt{-\gamma}\left(h_{ab} - \frac{1}{2}\gamma_{ab}\,\gamma^{cd}\, h_{cd}\right)\ .\end{aligned}$$

Using this equation, we then find

$$\gamma_{ab} = \frac{2h_{ab}}{\gamma^{cd}\, h_{cd}}\ , \qquad \gamma = \frac{4\det_{a,b}\left(h_{ab}\right)}{\left(\gamma^{cd}\, h_{cd}\right)^2}$$

and hence

$$\sqrt{-\gamma}\,\gamma^{ab}\, h_{ab} = \frac{2}{\gamma^{cd}\, h_{cd}}\sqrt{-\det_{a,b}\left(h_{ab}\right)}\,\gamma^{cd}\, h_{cd} = 2\sqrt{-\det_{a,b}\left(h_{ab}\right)}\ .$$

Chapter 3

3.1.

(a)

$$\begin{aligned} H &= \frac{1}{\pi} \int_0^\pi \mathrm{d}\sigma \, \Big(T_{++}(\tau,\sigma) + T_{--}(\tau,\sigma) \Big) \\ &= \frac{1}{\pi} \sum_{n=-\infty}^{\infty} \int_0^\pi \mathrm{d}\sigma \, \Big(\tilde{L}_n \, \mathrm{e}^{\,2\mathrm{i}n(\tau+\sigma)} + L_n \, \mathrm{e}^{\,2\mathrm{i}n(\tau-\sigma)} \Big) \\ &= \frac{1}{\pi} \sum_{n=-\infty}^{\infty} \delta_{n,0} \, \Big(\tilde{L}_n + L_n \Big) \, \mathrm{e}^{\,2\mathrm{i}n\tau} \\ &= \tilde{L}_0 + L_0 \ . \end{aligned}$$

Note that this result is independent of the worldsheet time τ, as it should be.

(b)

The canonically conjugate momentum is given by

$$\begin{aligned} p_\mu &= \pi \, \frac{\delta S[x, \mathrm{e}^{\,\phi} \, \eta]}{\delta \dot{x}^\mu} \\ &= \frac{1}{\alpha'} \left(2\alpha' \, p_0^\mu + \sqrt{2\alpha'} \, \sum_{n\neq 0} \left(\tilde{\alpha}_n^\mu \, \mathrm{e}^{\,-2\mathrm{i}n\sigma} + \alpha_n^\mu \, \mathrm{e}^{\,2\mathrm{i}n\sigma} \right) \, \mathrm{e}^{\,-2\mathrm{i}n\tau} \right) . \end{aligned}$$

The canonical equal-time Poisson brackets are

$$\begin{aligned} \big\{ x^\mu(\tau,\sigma) \, , \, x^\nu(\tau,\sigma'\,) \big\}_{\mathrm{PB}} &= 0 \ = \ \big\{ p^\mu(\tau,\sigma) \, , \, p^\nu(\tau,\sigma'\,) \big\}_{\mathrm{PB}} \ , \\ \big\{ x^\mu(\tau,\sigma) \, , \, p^\nu(\tau,\sigma'\,) \big\}_{\mathrm{PB}} &= \eta^{\mu\nu} \ \delta(\sigma - \sigma'\,) \ . \end{aligned}$$

Now use the Fourier series expansion

$$\delta(\sigma - \sigma'\,) = \frac{1}{\pi} \sum_{n=-\infty}^{\infty} \mathrm{e}^{\,2\mathrm{i}n\,(\sigma-\sigma'\,)} \ , \qquad \sigma, \sigma' \in [0, \pi] \ ,$$

and equate the Fourier modes on both sides of the above Poisson brackets, requiring that they be independent of τ.

3.2.

(a)

The angular momentum operators are given by

$$J^{\mu\nu} = \frac{1}{\pi} \int_0^\pi \mathrm{d}\sigma \, \Big(x^\mu(\tau,\sigma)\, p^\nu(\tau,\sigma) - x^\nu(\tau,\sigma)\, p^\mu(\tau,\sigma) \Big)$$

with

$$p^\mu(\tau,\sigma) = 2p_0^\mu + \sqrt{\frac{2}{\alpha'}} \sum_{n\neq 0} \alpha_n^\mu \sin(n\,\sigma) \ .$$

Using the orthogonality relations

$$\frac{1}{\pi} \int_0^\pi \mathrm{d}\sigma \ \mathrm{e}^{\pm\,\mathrm{i}\,n\,\sigma} = \delta_{n,0} \ ,$$

the result now easily follows.

(b)

Use the canonical commutators together with the Leibniz rules

$$[A\,B, C] = A\,[B, C] + [A, C]\,B \ ,$$

and so on. For example, since $[p_0^\mu, \alpha_n^\nu] = 0$ for $n \neq 0$, we compute

$$\begin{aligned} [p_0^\mu \, , \, J^{\nu\rho}] &= [p_0^\mu \, , \, x_0^\nu \, p_0^\rho] - [p_0^\mu \, , \, x_0^\rho \, p_0^\nu] \\ &= -\mathrm{i}\,\eta^{\mu\nu}\, p_0^\rho + \mathrm{i}\,\eta^{\mu\rho}\, p_0^\nu \ , \end{aligned}$$

where in the last line we have used $[p_0^\mu, p_0^\nu] = 0$ and $[x_0^\mu, p_0^\nu] = \mathrm{i}\,\eta^{\mu\nu}$.

(c)

Use

$$L_n = \frac{1}{2} \sum_{m=-\infty}^{\infty} \alpha_{n-m}^\rho \, \alpha_m^\lambda \, \eta_{\rho\lambda} \ , \qquad [\alpha_n^\mu \, , \, \alpha_m^\nu] = n\, \delta_{n+m,0}\, \eta^{\mu\nu} \ ,$$

and

$$\begin{aligned} &[\alpha_{-k}^\mu \, \alpha_k^\nu \, , \, \alpha_{n-m}^\rho \, \alpha_m^\lambda] \\ &\quad = k\,\eta^{\nu\rho}\,\delta_{k+n-m,0}\,\alpha_{-k}^\mu\,\alpha_m^\lambda + k\,\eta^{\nu\lambda}\,\delta_{k+m,0}\,\alpha_{-k}^\mu\,\alpha_{n-m}^\rho \\ &\qquad - k\,\eta^{\mu\rho}\,\delta_{n-m-k,0}\,\alpha_m^\lambda\,\alpha_k^\nu - k\,\eta^{\mu\lambda}\,\delta_{m-k,0}\,\alpha_k^\nu\,\alpha_{n-m}^\rho \ . \end{aligned}$$

The result now follows by noting that this commutator is invariant under the replacements $\mu \leftrightarrow \nu$ and $k \to -k$, and since $J^{\mu\nu} = -J^{\nu\mu}$.

3.3.

For $m \neq -n$, we have

$$[L_n, L_m] = \frac{1}{4} \sum_{k,l=-\infty}^{\infty} \eta_{\mu\nu}\,\eta_{\lambda\rho}\,\left[\alpha^{\mu}_{n-k}\,\alpha^{\nu}_{k}\,,\,\alpha^{\lambda}_{m-l}\,\alpha^{\rho}_{l}\right]$$

$$= \frac{1}{2} \sum_{k=-\infty}^{\infty} \eta_{\mu\nu}\left(k\,\alpha^{\mu}_{n-k}\,\alpha^{\nu}_{k+m} + (n-k)\,\alpha^{\mu}_{n+m-k}\,\alpha^{\nu}_{k}\right) .$$

Shifting $k \to k-m$ in the first sum thus shows that

$$[L_n, L_m] = (n-m)\,L_{n+m} \qquad \text{for} \quad m \neq -n \ .$$

For $m = -n$, we use the normal ordering prescription and $\zeta(-1) = -\frac{1}{12}$ to get the central term.

3.4.

(a)

We have

$$L_{-1}|k;0\rangle = \frac{1}{2} \sum_{m=-\infty}^{\infty} \alpha_{-1-m}\cdot\alpha_m|k;0\rangle$$

with $\alpha_m|k;0\rangle = 0$ for all $m > 0$ and $\alpha_{-1-m}|k;0\rangle = 0$ for all $m < -1$. Hence only the $m = 0, -1$ terms contribute to the sum, with equal values. Since $\alpha^{\mu}_0|k;0\rangle = \sqrt{2\alpha'}\ k^{\mu}\,|k;0\rangle$, we get

$$L_{-1}|k;0\rangle = \sqrt{2\alpha'}\ k\cdot\alpha_{-1}|k;0\rangle \ .$$

This is just the level 1 state $\sqrt{2\alpha'}\ |k;\zeta\rangle$ with polarization vector $\zeta = k$.

(b)

Physical states $|\text{phys}\rangle$ obey $L_1|\text{phys}\rangle = 0$, so

$$\begin{aligned}\langle\psi|\text{phys}\rangle &= \langle k;0|(L_{-1})^\dagger|\text{phys}\rangle \\ &= \langle k;0|L_1|\text{phys}\rangle \\ &= 0\,,\end{aligned}$$

where in the second line we used $\alpha_{-n} = (\alpha_n)^\dagger$, implying $L_{-n} = (L_n)^\dagger$.

(c)

$$\begin{aligned}L_1|\psi\rangle &= L_1\,L_{-1}|k;0\rangle \\ &= (L_{-1}\,L_1 + 2L_0)|k;0\rangle \\ &= \left(2\alpha'\,(p_0)^2\right)|k;0\rangle \\ &= 2\alpha'\,k^2\,|k;0\rangle\ .\end{aligned}$$

In the second line we have used the Virasoro algebra of Exercise 3.3.

Chapter 4

4.1.

(a)

By direct calculation using the Leibniz rule, we compute

$$\begin{aligned}\delta_\epsilon\big(\partial_a x^\mu\,\partial^a x_\mu\big) &= 2\partial_a x^\mu\,\partial^a(\delta_\epsilon x_\mu)\\ &= 2\,\overline{\epsilon}\,\partial_a x^\mu\,\partial^a\psi_\mu\ ,\end{aligned}$$

and

$$\begin{aligned}\delta_\epsilon\big(-\mathrm{i}\,\overline{\psi}^{\,\mu}\,\rho^a\,\partial_a\psi_\mu\big) &= -\mathrm{i}\,\big(\mathrm{i}\,\overline{\epsilon}\,\rho^a\,\partial_a x^\mu\,\rho^b\,\partial_b\psi_\mu - \mathrm{i}\,\overline{\psi}^{\,\mu}\,\rho^b\,\partial_b\rho^a\,\partial_a x_\mu\,\epsilon\big)\\ &= -2\,\overline{\epsilon}\,\rho^a\,\rho^b\,\partial_a\partial_b x^\mu\,\psi_\mu + \dots\\ &= -2\,\overline{\epsilon}\,\frac{1}{2}\,\big\{\rho^a\,,\,\rho^b\big\}\,\partial_a\partial_b x^\mu\,\psi_\mu + \dots\\ &= 2\,\overline{\epsilon}\,\partial_a\partial^a x^\mu\,\psi_\mu + \dots\\ &= -2\,\overline{\epsilon}\,\partial_a x^\mu\,\partial^a\psi_\mu + \dots\\ &= -\delta_\epsilon\big(\partial_a x^\mu\,\partial^a x_\mu\big) + \dots\ ,\end{aligned}$$

where throughout an ellipsis denotes total derivative terms.

(b)

Using the usual Noether prescription, we promote the global supersymmetry transformations of (a) above to local ones, *i.e.*, we allow non-constant $\epsilon = \epsilon(\xi)$. Consider the local supersymmetry variation of the Dirac Lagrangian, dropping terms proportional to $\epsilon(\xi)$ using the equations of motion to get

$$\begin{aligned}\delta_\epsilon\big(-\mathrm{i}\,\overline{\psi}^{\,\mu}\,\rho^a\,\partial_a\psi_\mu\big) &= -\mathrm{i}\,\big(\mathrm{i}\,\overline{\epsilon}\,\rho^a\,\partial_a x^\mu\,\rho^b\,\partial_b\psi_\mu - \mathrm{i}\,\overline{\psi}^{\,\mu}\,\rho^a\,\partial_a\rho^b\,\partial_b x_\mu\,\epsilon\big)\\ &= -2\,\partial_b\overline{\epsilon}\,\rho^a\,\rho^b\,\psi_\mu\,\partial_b x^\mu + \dots\ ,\end{aligned}$$

where here the ellipsis denotes terms proportional to the equations of motion. A similar calculation using the worldsheet Dirac algebra works for the bosonic $(\partial x)^2$ Lagrangian. Hence computing $\frac{\delta S}{\delta(\partial_a\overline{\epsilon})}$ yields the current J_a as required, where the normalization ensures that the associated quantum operator generates the correct supersymmetry transformations by taking graded commutators. Note that $\partial^a J_a = 0$ by the equations of motion.

(c)

$$\begin{aligned}
\rho^a\,J_a &= \frac{1}{2}\,\rho^a\,\rho^b\,\rho_a\,\psi^\mu\,\partial_b x_\mu \\
&= \frac{1}{2}\,\rho^a\Big(-2\,\delta^b{}_a - \rho_a\,\rho^b\Big)\,\psi^\mu\,\partial_b x_\mu \\
&= \frac{1}{2}\Big(-2\rho^b - \frac{1}{2}\,\{\rho^a\,,\,\rho_a\}\,\rho^b\Big)\,\psi^\mu\,\partial_b x_\mu \\
&= \frac{1}{2}\Big(-2\rho^b - \frac{1}{2}\,(-4)\,\rho^b\Big)\,\psi^\mu\,\partial_b x_\mu \\
&= 0\ .
\end{aligned}$$

4.2.

(a)

Writing the Dirac Lagrangian as

$$\frac{\mathrm{i}\,T}{2}\,\Big((\psi^\mu)^\dagger\,\dot\psi_\mu + \overline{\psi}{}^\mu\,\rho^1\,\psi'_\mu\Big)$$

gives the canonically conjugate fermionic momentum

$$P^\mu = 2\pi\,\alpha'\,\frac{\delta S}{\delta\dot\psi_\mu} = \frac{\mathrm{i}}{2}\,(\psi^\mu)^\dagger\ .$$

The non-vanishing canonical equal-time anticommutators are

$$\big\{P_\pm(\tau,\sigma)\,,\,\psi^\nu_\pm(\tau,\sigma'\,)\big\} = \mathrm{i}\,\eta^{\mu\nu}\,\delta(\sigma-\sigma'\,)\ .$$

Now expand $\delta(\sigma-\sigma'\,)$ in a Fourier series as before, and equate Fourier modes on both sides (requiring the left-hand side to again be τ-independent).

(b)

Use the oscillator commutators and anticommutators,

$$\big[\alpha^\mu_n\,,\,\alpha^\nu_m\big] = n\;\delta_{n+m,0}\;\eta^{\mu\nu}\ , \qquad \big\{\psi^\mu_r\,,\,\psi^\nu_s\big\} = \delta_{r+s,0}\;\eta^{\mu\nu}\ ,$$

to compute

$$[L_n, G_r] = \frac{1}{2} \sum_{m,m'=-\infty}^{\infty} \left[\alpha^\mu_{n-m}\, \alpha^\nu_m \,,\, \alpha^\rho_{m'}\right] \psi^\lambda_{r-m'}\, \eta_{\mu\nu}\, \eta_{\lambda\rho}$$

$$+ \frac{1}{4} \sum_{r'} \sum_{m'=-\infty}^{\infty} (2r' - n) \left[\psi^\mu_{n-r'}\, \psi^\nu_{r'} \,,\, \psi^\lambda_{r-m'}\right] \alpha^\rho_{m'}\, \eta_{\mu\nu}\, \eta_{\lambda\rho} \,.$$

The first term gives

$$\frac{1}{2} \sum_{m,m'=-\infty}^{\infty} \Big(\alpha^\mu_{n-m}\, \psi^\lambda_{r-m'}\, \delta_{m+m',0}\, \eta^{\nu\rho}\, \eta_{\mu\nu}\, \eta_{\lambda\rho}\, m$$

$$+ \alpha^\nu_m\, \psi^\lambda_{r-m'}\, \delta_{n-m+m',0}\, \eta^{\mu\rho}\, \eta_{\mu\nu}\, \eta_{\lambda\rho}\, (n-m)\Big)$$

$$= \sum_{m=-\infty}^{\infty} (n-m)\, \alpha_m \cdot \psi_{r+n-m} \,.$$

For the second term, we use

$$[A\,B, C] = A\,\{B, C\} - \{A, C\}\,B \,.$$

This yields

$$\frac{1}{4} \sum_{r'} \sum_{m'=-\infty}^{\infty} (2r' - n) \Big(\psi^\mu_{n-r'}\, \delta_{r+r'-m',0}\, \eta^{\nu\lambda}\, \eta_{\mu\nu}\, \eta_{\lambda\rho}$$

$$- \psi^\nu_{r'}\, \delta_{n-r'+r-m',0}\, \eta^{\mu\lambda}\, \eta_{\mu\nu}\, \eta_{\lambda\rho}\Big)\, \alpha^\rho_{m'}$$

$$= \sum_{m=-\infty}^{\infty} \left(m - r - \frac{n}{2}\right) \alpha_m \cdot \psi_{n-m+r} \,.$$

Adding these two terms together gives

$$[L_n, G_r] = \left(\frac{n}{2} - r\right) \sum_{m=-\infty}^{\infty} \alpha_m \cdot \psi_{n+r-m} = \left(\frac{n}{2} - r\right) G_{n+r} \,.$$

The anticommutator $\{G_r, G_s\}$ is computed similarly.

(c)

The fermionic part of the angular momentum operator is given by

$$K^{\mu\nu} = \frac{1}{\pi} \int_0^\pi \mathrm{d}\sigma \left((\psi^\mu)^\dagger\, P^\nu - (\psi^\nu)^\dagger\, P^\mu \right) ,$$

where the conjugate fermionic momentum is $P^\mu = \frac{\mathrm{i}}{2}\,\psi^\mu$. The result now follows easily from the mode expansion of ψ^μ and orthogonality of the Fourier basis over $\sigma \in [0, \pi]$.

4.3.

(a)

Using $[\alpha_n, \alpha_m] = 0$ for $n \neq -m$ and $a_{-n}\, a_n = 0, 1, 2, 3, \ldots$, we compute

$$\begin{aligned} \mathrm{tr}\left(q^{\sum_{n=1}^{\infty} \alpha_{-n}\,\alpha_n}\right) &= \prod_{n=1}^{\infty} \mathrm{tr}\left(q^{\alpha_{-n}\,\alpha_n}\right) \\ &= \prod_{n=1}^{\infty} \mathrm{tr}\left(q^{n\, a_{-n}\, a_n}\right) \\ &= \prod_{n=1}^{\infty} \sum_{m=0}^{\infty} \left(q^n\right)^m \\ &= \prod_{n=1}^{\infty} \frac{1}{1-q^n}\ . \end{aligned}$$

(b)

Using $\lambda_{-n}\,\lambda_n = 0, 1$ by Fermi statistics, we compute

$$\begin{aligned} \mathrm{tr}\left(q^{\sum_{n=1}^{\infty} \psi_{-n}\,\psi_n}\right) &= \prod_{n=1}^{\infty} \mathrm{tr}\left(q^{\psi_{-n}\,\psi_n}\right) \\ &= \prod_{n=1}^{\infty} \mathrm{tr}\left(q^{n\,\lambda_{-n}\,\lambda_n}\right) \\ &= \prod_{n=1}^{\infty} \left(1+q^n\right)\ . \end{aligned}$$

4.4.

Under the generating transformation $T : \tau \mapsto \tau' = \tau + 1$, so $\tau_2' = \tau_2$ and hence

$$\frac{\left|\eta(\tau')\right|^{-48}}{\tau_2'^{\;12}} = \frac{\left|\eta(\tau)\right|^{-48}}{\tau_2^{12}}\ .$$

Under $S : \tau \mapsto \tau' = -\frac{1}{\tau}$, so $\tau_2' = \frac{\tau_2}{|\tau|^2}$ and $|\eta(\tau')|^{-48} = |\tau|^{-24}\,|\eta(\tau)|^{-48}$ giving

$$\frac{\left|\eta(\tau')\right|^{-48}}{\tau_2'^{\;12}} = \frac{|\tau|^{24}}{\tau_2^{12}}\,|\tau|^{-24}\left|\eta(\tau)\right|^{-48} = \frac{\left|\eta(\tau)\right|^{-48}}{\tau_2^{12}}\,.$$

Chapter 5

5.1.

(a)

Look at the $n = 1$ case first to get a feel for what's going on. Since $\Gamma^\mu \Gamma^\nu = -\Gamma^\nu \Gamma^\mu$ for $\mu \neq \nu$ and $(\Gamma^\mu)^2 = 1$, we find

$$\begin{aligned}\Gamma^{11} \Gamma^\mu &= \Gamma^0 \Gamma^1 \cdots \Gamma^{\mu-1} \Gamma^\mu \Gamma^{\mu+1} \cdots \Gamma^9 \Gamma^\mu \\ &= \Gamma^0 \Gamma^1 \cdots \Gamma^{\mu-1} \Gamma^{\mu+1} \cdots \Gamma^9 \\ &= \frac{1}{9!} \epsilon^\mu{}_{\nu_1 \cdots \nu_9} \Gamma^{\nu_1} \cdots \Gamma^{\nu_9} \\ &= \frac{1}{(9!)^2} \epsilon^\mu{}_{\nu_1 \cdots \nu_9} \Gamma^{[\nu_1} \cdots \Gamma^{\nu_9]} \; .\end{aligned}$$

In the general case, we iterate the $n = 1$ case and cycle the Γ^{μ_i}'s through the Γ^ν's to get

$$\begin{aligned}\Gamma^{11} \Gamma^{[\mu_1} \cdots \Gamma^{\mu_n]} &= \sum_{\pi \in S_n} (-1)^\pi \Gamma^{11} \Gamma^{\mu_{\pi_1}} \cdots \Gamma^{\mu_{\pi_n}} \\ &= (-1)^{[\frac{n}{2}]} \sum_{\pi \in S_n} (-1)^\pi \prod_{\substack{\lambda=0 \\ \lambda \neq \mu_{\pi_1}, \ldots, \mu_{\pi_n}}}^{9} \Gamma^\lambda \\ &= \frac{(-1)^{[\frac{n}{2}]}}{(10-n)!} \sum_{\pi \in S_n} (-1)^\pi \epsilon^{\mu_{\pi_1} \cdots \mu_{\pi_n}}{}_{\nu_1 \cdots \nu_{10-n}} \Gamma^{\nu_1} \cdots \Gamma^{\nu_{10-n}} \\ &= \frac{(-1)^{[\frac{n}{2}]} n!}{(10-n)!} \epsilon^{\mu_1 \cdots \mu_n}{}_{\nu_1 \cdots \nu_{10-n}} \Gamma^{\nu_1} \cdots \Gamma^{\nu_{10-n}} \\ &= \frac{(-1)^{[\frac{n}{2}]} n!}{\left((10-n)!\right)^2} \epsilon^{\mu_1 \cdots \mu_n}{}_{\nu_1 \cdots \nu_{10-n}} \Gamma^{[\nu_1} \cdots \Gamma^{\nu_{10-n}]} \; .\end{aligned}$$

The products $\Gamma^{[\mu_1} \cdots \Gamma^{\mu_n]} \Gamma^{11}$ are found similarly.

(b)

Again, we start with the $n = 1$ case and compute

$$\Gamma^\nu \Gamma^\mu = \frac{1}{2} \left[\Gamma^\nu , \Gamma^\mu\right] + \frac{1}{2} \left\{\Gamma^\nu , \Gamma^\mu\right\} = \frac{\Gamma^{[\nu} \Gamma^{\mu]}}{2} + \eta^{\nu\mu} \; .$$

Iterating this formula, the general case is then

$$\begin{aligned}
\Gamma^\nu\,\Gamma^{[\mu_1}\cdots\Gamma^{\mu_n]} &= \sum_{\pi\in S_n}(-1)^\pi\,\Gamma^\nu\,\Gamma^{\mu_{\pi 1}}\cdots\Gamma^{\mu_{\pi n}} \\
&= \sum_{\pi\in S_n}(-1)^\pi\left(\frac{1}{2}\,\Gamma^{[\nu}\,\Gamma^{\mu_{\pi 1}]}\,\Gamma^{\mu_{\pi 2}}\cdots\Gamma^{\mu_{\pi n}}\right. \\
&\qquad\left.+\,\eta^{\nu\mu_{\pi 1}}\,\Gamma^{\mu_{\pi 2}}\cdots\Gamma^{\mu_{\pi n}}\right) \\
&= \sum_{\pi\in S_n}(-1)^\pi\left(\frac{1}{(n+1)\,n!}\,\Gamma^{[\nu}\,\Gamma^{\mu_{\pi 1}}\cdots\Gamma^{\mu_{\pi n}]}\right. \\
&\qquad\left.+\,\frac{1}{\big((n-1)!\big)^2}\,\eta^{\nu[\mu_{\pi 1}}\,\Gamma^{\mu_{\pi 2}}\cdots\Gamma^{\mu_{\pi n}]}\right)\,,
\end{aligned}$$

and the result follows.

5.2.

(a)

For d even, a spinor in d dimensions has $2^{d/2}$ components. The Majorana and Weyl conditions each reduce this by $\frac{1}{2}$, so the number of Majorana–Weyl components is

$$\frac{1}{2}\times\frac{1}{2}\times 2^{d/2} = 2^{d/2-2} = 2^3 = 8$$

for $d = 10$. Hence the tensor product has $8\times 8 = 64$ independent components. On the other hand, the number of independent components of $F^{(n)}_{\mu_1\cdots\mu_n}$ is $\frac{1}{2}\,(d-n+1)\,(d-n+2)$ for n even, and summing this over $n = 2,4,6,8,10$ for $d = 10$ then yields the desired result.

(b)

By the Dirac–Ramond equation and Exercise 5.1(b), we have

$$\begin{aligned}
0 &= \alpha_0\cdot\psi_0|\psi_{\rm l}\rangle_{\rm R}\otimes|\psi_{\rm r}\rangle_{\rm R} \\
&= (p_0)_\nu\sum_n F^{(n)}_{\mu_1\cdots\mu_n}\,\Gamma^\nu\,\Gamma^{[\mu_1}\cdots\Gamma^{\mu_n]} \\
&= (p_0)_\nu\sum_n F^{(n)}_{\mu_1\cdots\mu_n}\left(\frac{\Gamma^{[\nu}\,\Gamma^{\mu_1}\cdots\Gamma^{\mu_n]}}{n+1}\right. \\
&\qquad\left.+\,\frac{n}{(n-1)!}\,\eta^{\nu[\mu_1}\,\Gamma^{\mu_2}\cdots\Gamma^{\mu_n]}\right)\,.
\end{aligned}$$

Now use completeness of the antisymmetrized Dirac matrix products to deduce that both sets of terms vanish, for each n. The first one yields

$$(p_0)_{[\nu}\, F^{(n)}_{\mu_1\cdots\mu_n]} = 0 \ ,$$

while the second one gives

$$(p_0)_\nu\, \eta^{\nu\mu_1}\, F^{(n)}_{\mu_1\cdots\mu_n} = 0 \ .$$

With $(p_0)_\mu = -\mathrm{i}\,\partial_\mu$, we get the desired result.

5.3.

Write the σ and τ derivatives in light-cone coordinates

$$\partial_\sigma = \partial_+ - \partial_- \ , \qquad \partial_\tau = \partial_+ + \partial_- \ .$$

Then

$$\partial_\tau x = (\partial_+ + \partial_-)(x_{\mathrm{L}} + x_{\mathrm{R}}) = \partial_+ x_{\mathrm{L}} + \partial_- x_{\mathrm{R}} \ .$$

The T-duality transformation maps this last expression to

$$\partial_+ x_{\mathrm{L}} - \partial_- x_{\mathrm{R}} = (\partial_+ - \partial_-)(x_{\mathrm{L}} + x_{\mathrm{R}}) = \partial_\sigma x \ .$$

Hence T-duality interchanges $\partial_\tau \leftrightarrow \partial_\sigma$.

Chapter 6

6.1.

The canonical momentum is

$$p^\mu = \frac{\delta S}{\delta \dot{x}_\mu} = m\,\dot{x}^\mu - q\,A^\mu = p^\mu_{\rm mech} - q\,A^\mu$$

where $p^\mu_{\rm mech} = m\,\dot{x}^\mu$ is the mechanical momentum. Note that only the canonical momentum is gauge-invariant. Thus

$$p^d = p^d_{\rm mech} - \frac{q\,\theta}{2\pi\,R} \ .$$

As usual, $p^d_{\rm mech} = \frac{n}{R}$ must be quantized as before, and so

$$p^d = \frac{n}{R} - \frac{q\,\theta}{2\pi\,R}$$

as required.

6.2.

(a)

We have

$$\partial_{\overline{z}}\,\frac{1}{z-z'} = 2\pi\,\delta(z-z') \ ,$$

because

$$\partial_{\overline{z}}\,\frac{1}{z-z'} = 0 \ , \qquad z \neq z'$$

and

$$\int {\rm d}z\ {\rm d}\overline{z}\ \partial_{\overline{z}}\,\frac{1}{z-z'} = \int \frac{{\rm d}z}{z-z'} = 2\pi$$

by the Cauchy integral formula. Thus the Green's function on $\mathbb{C}$ is given by $\frac{1}{2\pi}\,\ln(z-z')$. We can satisfy the Neumann boundary conditions across the real line $\mathbb{R} \subset \mathbb{C}$ by using the method of images. We place a charge at a point z' in the upper complex half-plane. Its image charge then sits at the point $\overline{z}\,'$ in the lower complex half-plane. Since $\overline{z}\,'^{\,-1}$ lives in the upper half-plane, we then find

$$N(z,z') = \frac{1}{2\pi}\,\ln(z-z') + \frac{1}{2\pi}\,\ln\Big(z - \frac{1}{\overline{z}\,'}\Big) = \frac{1}{2\pi}\,\ln\big(z-z'\big)\,\big(z-\overline{z}\,'^{\,-1}\big) \ .$$

Now add to this the complex conjugate for reality.

(b)

$$
\begin{aligned}
N(\mathrm{e}^{\mathrm{i}\theta}, \mathrm{e}^{\mathrm{i}\theta'}) &= \frac{1}{2\pi}\Big(\ln\big(\mathrm{e}^{\mathrm{i}\theta} - \mathrm{e}^{\mathrm{i}\theta'}\big) + \ln\big(\mathrm{e}^{-\mathrm{i}\theta} - \mathrm{e}^{-\mathrm{i}\theta'}\big)\Big) \\
&= \frac{1}{2\pi}\Big(\ln\big(1 - \mathrm{e}^{-\mathrm{i}(\theta-\theta')}\big) + \ln\big(1 - \mathrm{e}^{\mathrm{i}(\theta-\theta')}\big)\Big) \\
&= \frac{1}{2\pi}\Big(-\sum_{n=1}^{\infty}\frac{1}{n}\,\mathrm{e}^{-\mathrm{i}n(\theta-\theta')} - \sum_{n=1}^{\infty}\frac{1}{n}\,\mathrm{e}^{\mathrm{i}n(\theta-\theta')}\Big) \\
&= -\frac{1}{\pi}\sum_{n=1}^{\infty}\frac{\cos\big(n(\theta-\theta')\big)}{n} \ .
\end{aligned}
$$

6.3.

(a)

Supersymmetry of the bulk Lagrangian is as previously. The boundary supersymmetry transformations are

$$
\delta_\epsilon x^\mu = \epsilon\,\psi^\mu \ , \qquad \delta_\epsilon \psi^\mu = \epsilon\,\dot{x}^\mu \ .
$$

Using the chain rule $\delta_\epsilon F(x) = (\delta_\epsilon x^\mu)\,\partial_\mu F(x)$, we find

$$
\begin{aligned}
\delta_\epsilon\big(\dot{x}^\mu\,A_\mu(x)\big) &= \epsilon\,\dot{\psi}^\mu\,A_\mu(x) + \epsilon\,\dot{x}^\mu\,\psi^\nu\,\partial_\nu A_\mu(x) \\
&= \epsilon\,\big(-\dot{x}^\nu\,\psi^\mu\,\partial_\nu A_\mu(x) + \dot{x}^\mu\,\psi^\nu\,\partial_\nu A_\mu(x)\big) + \dots \\
&= -\epsilon\,\dot{x}^\mu\,\psi^\nu\,F_{\mu\nu}(x)
\end{aligned}
$$

with the ellipsis denoting total boundary derivative terms, and

$$
\begin{aligned}
\delta_\epsilon\big(\psi^\mu\,\psi^\nu\,F_{\mu\nu}(x)\big) &= F_{\mu\nu}(x)\,\big(\epsilon\,\dot{x}^\mu\,\psi^\nu - \epsilon\,\dot{x}^\nu\,\psi^\mu\big) \\
&\quad + \psi^\mu\,\psi^\nu\,\epsilon\,\psi^\lambda\,\partial_\lambda F_{\mu\nu}(x) \\
&= 2\epsilon\,F_{\mu\nu}(x)\,\dot{x}^\mu\,\psi^\nu \\
&\quad + \psi^\mu\,\psi^\nu\,\psi^\lambda\,\big(\partial_\lambda\partial_\mu A_\nu(x) - \partial_\lambda\partial_\nu A_\mu(x)\big) \ .
\end{aligned}
$$

The second term in the last line here vanishes, since $\psi^\mu\,\psi^\lambda$ is antisymmetric while $\partial_\mu\partial_\lambda$ is symmetric in μ, λ, so

$$
\delta_\epsilon\big(\psi^\mu\,\psi^\nu\,F_{\mu\nu}(x)\big) = -2\delta_\epsilon\big(\dot{x}^\mu\,A_\mu(x)\big)
$$

as required.

(b)

As before,

$$\partial_{\overline{z}} \frac{1}{z - z'} = 2\pi \, \delta(z - z') \ ,$$

so the fermionic Green's function on $\mathbb{C}$ is $\frac{1}{2\pi} \frac{1}{z-z'}$. Since the differential operator ∂_z is antihermitean, using the method of images again we get

$$K(z, z') = \frac{1}{2\pi} \left(\frac{1}{z - z'} - \frac{1}{z - \overline{z}'^{\,-1}} \right)$$

and hence

$$\begin{aligned} K(\mathrm{e}^{\mathrm{i}\,\theta}, \mathrm{e}^{\mathrm{i}\,\theta'}) &= \frac{1}{2\pi} \, \mathrm{e}^{-\mathrm{i}\,\theta} \left(\frac{1}{1 - \mathrm{e}^{-\mathrm{i}\,(\theta - \theta')}} - \frac{1}{1 - \mathrm{e}^{\mathrm{i}\,(\theta - \theta')}} \right) \\ &= \frac{1}{2\pi} \, \mathrm{e}^{-\mathrm{i}\,\theta} \left(\sum_{r \geq 0} \mathrm{e}^{-\mathrm{i}\,r\,(\theta - \theta')} - \sum_{r \geq 0} \mathrm{e}^{\mathrm{i}\,r\,(\theta - \theta')} \right) \\ &= -\frac{\mathrm{i}\,\mathrm{e}^{-\mathrm{i}\,\theta}}{\pi} \sum_{r \geq 0} \sin\left(r\,(\theta - \theta')\right) . \end{aligned}$$

Now take the modulus for reality, and restrict $r = \frac{1}{2}, \frac{3}{2}, \ldots$ to half-integer values to get the required anti-periodicity under $\theta \to \theta + 2\pi$.

(c)

The bosonic contribution is as before. It multiplies the fermionic contribution, obtained by integrating out bulk modes from the boundary action

$$S_{\text{ferm}} = \frac{1}{2} \int_0^{2\pi} \mathrm{d}\theta \left(\frac{1}{2\pi\,\alpha'} \, \psi^\mu \, K^{-1} \, \psi_\mu + \mathrm{i}\, F_{\mu\nu} \, \psi^\mu \, \psi^\nu \right) ,$$

where

$$K^{-1}(\theta, \theta') = -\frac{1}{\pi} \sum_r \sin\left(r\,(\theta - \theta')\right) .$$

On the boundary, we expand the anti-periodic fermion fields as

$$\psi^\mu(\theta) = \sum_r \chi_r^\mu \, \sin(r\,\theta) \ .$$

Note that anti-periodic fermions have no zero modes. The pertinent Gaussian forms are now given by

$$\frac{1}{2}\,\frac{1}{2\pi\,\alpha'}\,\frac{1}{2\pi\,r^2}\,\left(\chi_r^{2l-1}\quad,\quad\chi_r^{2l}\right)\begin{pmatrix}1 & -2\pi\,\alpha'\,f_l\\ 2\pi\,\alpha'\,f_l & 1\end{pmatrix}\begin{pmatrix}\chi_r^{2l-1}\\ \chi_r^{2l}\end{pmatrix}\,.$$

The $\frac{1}{r^2}$ here arises because K^{-1} has no extra factors of r (like N^{-1} did). The modes χ_r^μ are real Grassmann variables, so now Gaussian integration puts the fluctuation determinants upstairs with

$$\begin{aligned}
&\prod_r\left\{\Big(\frac{1}{4\pi^2\,\alpha'\,r^2}\Big)^2\left[1+(2\pi\,\alpha'\,f_l)^2\right]\right\}\\
&\quad=\Big(\prod_{n=0}^{\infty}\frac{1}{\left(n+\frac{1}{2}\right)^2}\Big)\left\{\Big(\frac{1}{4\pi^2\,\alpha'}\Big)^2\left[1+(2\pi\,\alpha'\,f_l)^2\right]\right\}^{\zeta(0,1/2)}\\
&\quad=\prod_{n=0}^{\infty}\frac{1}{\left(n+\frac{1}{2}\right)^2}\,,
\end{aligned}$$

where we have used $\zeta(0,1/2)=0$. For large n, this cancels the divergent product $\prod_{n=1}^{\infty}n^2$ arising from the bosonic integration.

Chapter 7

7.1.

(a)

Varying the Born–Infeld action gives

$$\begin{aligned}
0 &= \frac{\delta S_{\rm BI}}{\delta A_\mu} \\
&= \frac{T^{-1}}{2}\left(\frac{1}{\mathbb{1}+T^{-1}\,F}\right)^{\nu\lambda}\left(\frac{\delta}{\delta A_\mu}F_{\lambda\nu}\right)\,\det{}^{1/2}(\mathbb{1}+T^{-1}\,F) \\
&= -\frac{T^{-1}}{2}\left(\frac{1}{\mathbb{1}+T^{-1}\,F}\right)^{\nu\rho}(\partial_\nu F_{\lambda\mu})\left(\frac{1}{\mathbb{1}+T^{-1}\,F}\right)^{\lambda\rho} \\
&\qquad \times\det{}^{1/2}(\mathbb{1}+T^{-1}\,F)+\dots
\end{aligned}$$

after an integration by parts, where the ellipsis denotes total derivative terms. Now note that $F^\top = -F$ implies

$$(\mathbb{1}+F)\,(\mathbb{1}+F)^\top = (\mathbb{1}+F)\,(\mathbb{1}-F) = \mathbb{1}-F^2\ ,$$

so that

$$0 = \left(\frac{1}{\mathbb{1}-(T^{-1}\,F)^2}\right)^{\nu\lambda}(\partial_\nu F_{\lambda\mu})$$

as required.

(b)

Rotate to a coordinate system in which the 4×4 antisymmetric matrix $(F_{\mu\nu})$ is skew-diagonal, i.e. in its Jordan canonical form

$$(F_{\mu\nu}) = \begin{pmatrix} 0 & -f_1 & 0 & 0 \\ f_1 & 0 & 0 & 0 \\ 0 & 0 & 0 & -f_2 \\ 0 & 0 & f_2 & 0 \end{pmatrix} .$$

Then

$$\det\left(\mathbb{1}+T^{-1}F\right)=\begin{vmatrix}1 & -T^{-1}f_1 & 0 & 0\\ T^{-1}f_1 & 1 & 0 & 0\\ 0 & 0 & 1 & -T^{-1}f_2\\ 0 & 0 & T^{-1}f_2 & 1\end{vmatrix}$$
$$=\left(1+T^{-2}f_1^2\right)\left(1+T^{-2}f_2^2\right) .$$

The matrix $(\tilde{F}_{\mu\nu})$ is given by

$$(\tilde{F}_{\mu\nu})=\begin{pmatrix}0 & -f_2 & 0 & 0\\ f_2 & 0 & 0 & 0\\ 0 & 0 & 0 & -f_1\\ 0 & 0 & f_1 & 0\end{pmatrix},$$

so that

$$F_{\mu\nu}\,F^{\mu\nu}=-\operatorname{Tr}\left(F^2\right)\ =\ 2f_1^2+2f_2^2\ ,$$
$$F_{\mu\nu}\,\tilde{F}^{\mu\nu}=-\operatorname{Tr}\left(F\,\tilde{F}\right)\ =\ 4f_1\,f_2\ .$$

Thus

$$\begin{aligned}1+\frac{1}{2T^2}\,F_{\mu\nu}\,F^{\mu\nu}-\frac{1}{16T^4}\left(F_{\mu\nu}\,\tilde{F}^{\mu\nu}\right)^2&\\ =\ &1+T^{-2}\left(f_1^2+f_2^2\right)-T^{-4}\,f_1^2\,f_2^2\\ =\ &\left(1+T^{-2}\,f_1^2\right)\left(1+T^{-2}\,f_2^2\right)\\ =\ &\det\left(\mathbb{1}+T^{-1}\,F\right) .\end{aligned}$$

(c)

Take $F_{rt}=E_r\neq 0$ and all other components $F_{\mu\nu}=0$. Then

$$F_{\mu\nu}\,F^{\lambda\nu}=E_r^2\,\delta_\mu{}^\lambda\ ,$$

and hence

$$\begin{aligned}\left(\frac{1}{\mathbb{1}-\left(T^{-1}\,F\right)^2}\right)^{\nu\lambda}\partial_\nu F_{\lambda\mu}&=\frac{1}{1-T^{-2}\,E_r^2}\,\eta^{\nu\lambda}\,\partial_\nu F_{\lambda\mu}\\ &=\frac{\nabla_r E_r}{1-T^{-2}\,E_r^2}\,\delta_{\mu,t}\\ &=j^t\,\delta_{\mu,t}\ .\end{aligned}$$

For a point charge Q at $\vec{x} = \vec{0}$ we put $j^t(\vec{x}\,) = Q\ \delta^{(3)}(\vec{x}\,)$. Rescaling $E_r \to T^{-1}\,E_r$, we then get the equation

$$\nabla_r E_r = 4\pi^2\,r_0^2\,\frac{\delta(r)}{2r}\,\left(1 - E_r^2\right)\,.$$

Substituting in the ansatz

$$E_r = \frac{Q}{\sqrt{r^4 + r_0^4}}$$

and integrating this differential equation over all of three-space shows that this is indeed the required solution here.

7.2.

We have

$$\begin{aligned}\begin{vmatrix} \mathcal{N} & -\mathcal{A}^\top \\ \mathcal{A} & \mathcal{M} \end{vmatrix} &= \begin{vmatrix} \mathbb{1}_q & 0_{q\times p} \\ 0_{p\times q} & \mathcal{M} \end{vmatrix}\,\begin{vmatrix} \mathcal{N} & -\mathcal{A}^\top \\ \mathcal{M}^{-1}\,\mathcal{A} & \mathbb{1}_p \end{vmatrix} \\ &= \det(\mathcal{M})\,\begin{vmatrix} \mathcal{N} & -\mathcal{A}^\top \\ \mathcal{M}^{-1}\,\mathcal{A} & \mathbb{1}_p \end{vmatrix}\,.\end{aligned}$$

Now evaluate the second determinant by performing a minor expansion along the last column to find that it evaluates just like a 2×2 determinant. Thus

$$\begin{vmatrix} \mathcal{N} & -\mathcal{A}^\top \\ \mathcal{A} & \mathcal{M} \end{vmatrix} = \det(\mathcal{M})\,\det\left(\mathcal{N} + \mathcal{A}^\top\,\mathcal{M}^{-1}\,\mathcal{A}\right)\,.$$

The other identity is similarly found.

7.3.

We have

$$\begin{aligned}\delta_\epsilon\left(F_{\mu\nu}\,F^{\mu\nu}\right) &= 2F_{\mu\nu}\,\delta_\epsilon\left(F^{\mu\nu}\right) \\ &= \mathrm{i}\,\bar\epsilon\,F_{\mu\nu}\,\left(\Gamma^\nu\,\partial^\mu\psi - \Gamma^\mu\,\partial^\nu\psi\right) \\ &= 2\,\mathrm{i}\,\bar\epsilon\,F_{\mu\nu}\,\Gamma^\nu\,\partial^\mu\psi\end{aligned}$$

and

$$
\begin{aligned}
&\delta_\epsilon\big(\overline{\psi}\,\Gamma^\mu\,(\partial_\mu\psi - \mathrm{i}\,[A_\mu,\psi])\big) \\
&\quad = -\frac{1}{2}\,\overline{\epsilon}\,F_{\mu\nu}\,\left[\Gamma^\mu\,,\,\Gamma^\nu\right]\Gamma^\lambda\,(\partial_\lambda\psi - \mathrm{i}\,[A_\lambda,\psi]) \\
&\qquad - \frac{1}{2}\,\overline{\psi}\,\Gamma^\lambda\left[\Gamma^\mu\,,\,\Gamma^\nu\right]\epsilon\,\partial_\lambda F_{\mu\nu} \\
&\qquad + \frac{1}{2}\,\overline{\psi}\,\Gamma^\lambda\left[(\overline{\epsilon}\,\Gamma_\mu\,\psi)\,\psi - \psi\,(\overline{\epsilon}\,\Gamma_\mu\,\psi)\right] \\
&\qquad + \frac{\mathrm{i}}{2}\,\overline{\psi}\,\Gamma^\lambda\left[A_\lambda\,,\,F_{\mu\nu}\right]\left[\Gamma^\mu\,,\,\Gamma^\nu\right]\epsilon\;.
\end{aligned}
$$

The first term here vanishes due to the (covariant) Dirac equation. For the remaining terms, we use the Dirac algebra and Exercise 5.1(b) to get

$$
\Gamma^\lambda\left[\Gamma^\mu\,,\,\Gamma^\nu\right] = \frac{\Gamma^{[\lambda}\,\Gamma^\mu\,\Gamma^{\nu]}}{3} + 2\eta^{\lambda\mu}\,\Gamma^\nu - 2\eta^{\lambda\nu}\,\Gamma^\mu\;.
$$

Using further the Yang–Mills equations of motion, we finally arrive at

$$
\delta_\epsilon\big(\overline{\psi}\,\Gamma^\mu\,D_\mu\psi\big) = \frac{\mathrm{i}}{2}\,\delta_\epsilon\big(F_{\mu\nu}\,F^{\mu\nu}\big)
$$

as required.

Bibliography

A. Abouelsaood, C.G. Callan, C.R. Nappi and S.A. Yost, "Open Strings in Background Gauge Fields", Nucl. Phys. **B280** [FS18] (1987) 599–624.

O. Aharony, S.S. Gubser, J.M. Maldacena, H. Ooguri and Y. Oz, "Large N Field Theories, String Theory and Gravity", Phys. Rept. **323** (2000) 183–386 [`arXiv:hep-th/9905111`].

L. Alvarez-Gaumé and P. Ginsparg, "The Structure of Gauge and Gravitational Anomalies", Ann. Phys. **161** (1985) 423.

L. Alvarez-Gaumé and E. Witten, "Gravitational Anomalies", Nucl. Phys. **B234** (1984) 269.

D. Amati, M. Ciafaloni and G. Veneziano, "Can Spacetime be Probed Below the String Size?", Phys. Lett. **B216** (1989) 41.

J. Ambjørn, Y.M. Makeenko, G.W. Semenoff and R.J. Szabo, "String Theory in Electromagnetic Fields", J. High Energy Phys. **0302** (2003) 026 [`arXiv:hep-th/0012092`].

I. Antoniadis, N. Arkani-Hamed, S. Dimopoulos and G. Dvali, "New Dimensions at a Millimeter to a Fermi and Superstrings at a TeV", Phys. Lett. **B436** (1998) 257–263 [`arXiv:hep-ph/9804398`].

N. Arkani-Hamed, S. Dimopoulos and G. Dvali, "The Hierarchy Problem and New Dimensions at a Millimeter", Phys. Lett. **B429** (1998) 263–272 [`arXiv:hep-ph/9803315`].

T. Banks, W. Fischler, S.H. Shenker and L. Susskind, "M-Theory as a Matrix Model: A Conjecture", Phys. Rev. **D55** (1997) 5112–5128 [`arXiv:hep-th/9610043`].

A.A. Belavin, A.M. Polyakov and A.B. Zamolodchikov, "Infinite Conformal Symmetry in Two-Dimensional Quantum Field Theory", Nucl. Phys. **B241** (1984) 333–380.

N. Berkovits, "ICTP Lectures on Covariant Quantization of the Superstring", (2002) [`arXiv:hep-th/0209059`].

M. Born and L. Infeld, "Foundations of the New Field Theory", Proc. Roy. Soc. London **A144** (1934) 425–451.

R.L. Bryant and E. Sharpe, "D-Branes and spin^c Structures", Phys. Lett. **B450** (1999) 353–357 [`arXiv:hep-th/9812084`].

C.G. Callan and J.M. Maldacena, "D-Brane Approach to Black Hole Quantum Mechanics", Nucl. Phys. **B472** (1996) 591–610 [arXiv:hep-th/9602043].

P. Candelas, G.T. Horowitz, A. Strominger and E. Witten, "Vacuum Configurations for Superstrings", Nucl. Phys. **B258** (1985) 46–74.

H.-M. Chan and J.E. Paton, "Generalized Veneziano Model with Isospin", Nucl. Phys. **B10** (1969) 516–520.

Y.-K.E. Cheung and Z. Yin, "Anomalies, Branes and Currents", Nucl. Phys. **B517** (1998) 69–91 [arXiv:hep-th/9710206].

A. Connes, M.R. Douglas and A. Schwarz, "Noncommutative Geometry and Matrix Theory: Compactification on Tori", J. High Energy Phys. **9802** (1998) 003 [arXiv:hep-th/9711162].

E. Cremmer, B. Julia and J. Scherk, "Supergravity Theory in Eleven-Dimensions", Phys. Lett. **B76** (1978) 409–412.

J. Dai, R.G. Leigh and J. Polchinski, "New Connections Between String Theories", Mod. Phys. Lett. **A4** (1989) 2073–2083.

M.R. Douglas, "Branes within Branes", in: *Strings, Branes and Dualities*, eds. L. Baulieu, P. Di Francesco, M.R. Douglas, V.A. Kazakov, M. Picco and P. Windey (Kluwer Academic Publishers, Dordrecht, 1999), pp. 267–275 [arXiv:hep-th/9512077].

M.R. Douglas, D. Kabat, P. Pouliot and S.H. Shenker, "D-Branes and Short Distances in String Theory", Nucl. Phys. **B485** (1997) 85–127 [arXiv:hep-th/9608024].

M.J. Duff, "M-Theory (The Theory Formerly Known as Strings)", Int. J. Mod. Phys. **A11** (1996) 5623–5642 [arXiv:hep-th/9608117].

M.J. Duff, R.R. Khuri and J.X. Lu, "String Solitons", Phys. Rept. **259** (1995) 213–326 [arXiv:hep-th/9412184].

J.M. Figueroa-O'Farrill, "BUSSTEPP Lectures on Supersymmetry", (2001) [arXiv:hep-th/0109172].

A. Font, L. Ibanez, D. Lust and F. Quevedo, "Strong-Weak Coupling Duality and Nonperturbative Effects in String Theory", Phys. Lett. **B249** (1990) 35–43.

E.S. Fradkin and A.A. Tseytlin, "Nonlinear Electrodynamics from Quantized Strings", Phys. Lett. **B163** (1985) 295.

P. Ginsparg, "Applied Conformal Field Theory", in: *Fields, Strings and Critical Phenomena*, eds. E. Brézin and J. Zinn-Justin (North Holland, Amsterdam, 1990), pp. 1–168.

A. Giveon, M. Porrati and E. Rabinovici, "Target Space Duality in String Theory", Phys. Rept. **244** (1994) 77–202 [arXiv:hep-th/9401139].

F. Gliozzi, J. Scherk and D.I. Olive, "Supergravity and the Spinor Dual Model", Phys. Lett. **B65** (1976) 282.

T. Goto, "Relativistic Quantum Mechanics of One-Dimensional Mechanical Continuum and Subsidiary Condition of Dual Resonance Model", Prog. Theor. Phys. **46** (1971) 1560–1569.

I.S. Gradshteyn and I.M. Ryzhik, *Table of Integrals, Series and Products* (Academic Press, San Diego, 1980).

M.B. Green and J.H. Schwarz, "Supersymmetrical Dual String Theory", Nucl. Phys. **B181** (1981) 502–530.

M.B. Green and J.H. Schwarz, "Supersymmetrical Dual String Theory 3: Loops and Renormalization", Nucl. Phys. **B198** (1982) 441–460.

M.B. Green and J.H. Schwarz, "Anomaly Cancellation in Supersymmetric $D = 10$ Gauge Theory and Superstring Theory", Phys. Lett. **B149** (1984) 117–122.

M.B. Green, J.A. Harvey and G. Moore, "I-Brane Inflow and Anomalous Couplings on D-Branes", Class. Quant. Grav. **14** (1997) 47–52 [`arXiv:hep-th/9605033`].

M.B. Green, J.H. Schwarz and E. Witten, *Superstring Theory* (Cambridge University Press, 1987).

D.J. Gross and P.F. Mende, "String Theory Beyond the Planck Scale", Nucl. Phys. **B303** (1988) 407.

D.J. Gross, J.A. Harvey, E. Martinec and R. Rohm, "The Heterotic String", Phys. Rev. Lett. **54** (1985) 502–505.

P. Hořava, "Strings on Worldsheet Orbifolds", Nucl. Phys. **B327** (1989) 461.

G.T. Horowitz and A. Strominger, "Black Strings and p-Branes", Nucl. Phys. **B360** (1991) 197–209.

C.M. Hull and P.K. Townsend, "Unity of Superstring Dualities", Nucl. Phys. **B438** (1995) 109–137 [`arXiv:hep-th/9410167`].

C.V. Johnson, "D-Brane Primer", in: *Strings, Branes and Gravity*, eds. J.A. Harvey, S. Kachru and E. Silverstein (World Scientific, 2001), pp. 129–350 [`arXiv:hep-th/0007170`].

S. Kachru and C. Vafa, "Exact Results for $N = 2$ Compactifications of Heterotic Strings", Nucl. Phys. **B450** (1995) 69–89 [`arXiv:hep-th/9505105`].

R.G. Leigh, "Dirac–Born–Infeld Action from Dirichlet Sigma Model", Mod. Phys. Lett. **A4** (1989) 2767.

J.M. Maldacena, "The Large N Limit of Superconformal Field Theories and Supergravity", Adv. Theor. Math. Phys. **2** (1998) 231–252 [`arXiv:hep-th/9711200`].

N.E. Mavromatos and R.J. Szabo, "Matrix D-Brane Dynamics, Logarithmic Operators and Quantization of Noncommutative Spacetime", Phys. Rev. **D59** (1999) 104018 [`arXiv:hep-th/9808124`].

D. Mumford, *Tata Lectures on Theta* (Birkhäuser, Basel, 1983).

R.C. Myers, "Dielectric Branes", J. High Energy Phys. **9912** (1999) 022 [`arXiv:hep-th/9910053`].

Y. Nambu, "Strings, Monopoles and Gauge Fields", Phys. Rev. **D10** (1974) 4262.

A. Neveu and J.H. Schwarz, "Factorizable Dual Model of Pions", Nucl. Phys. **B31** (1971) 86–112.

J. Polchinski, "Dirichlet Branes and Ramond-Ramond Charges", Phys. Rev. Lett. **75** (1995) 4724–4727 [`arXiv:hep-th/9510017`].

J. Polchinski, *String Theory* (Cambridge University Press, 1998).

A.M. Polyakov, "Quantum Geometry of Bosonic Strings", Phys. Lett. **B103** (1981) 207–210.

A.M. Polyakov, "Quantum Geometry of Fermionic Strings", Phys. Lett. **B103** (1981) 211–213.

P. Ramond, "Dual Theory for Free Fermions", Phys. Rev. **D3** (1971) 2415–2418.

L. Randall and R. Sundrum, "A Large Mass Hierarchy from a Small Extra Dimension", Phys. Rev. Lett. **83** (1999) 3370–3373 [`arXiv:hep-ph/9905221`].

L. Randall and R. Sundrum, "An Alternative to Compactification", Phys. Rev. Lett. **83** (1999) 4690–4693 [`arXiv:hep-th/9906064`].

J. Scherk and J.H. Schwarz, "Dual Models for Non-Hadrons", Nucl. Phys. **B81** (1974) 118–144.

J.H. Schwarz, "Evidence for Nonperturbative String Symmetries", Lett. Math. Phys. **34** (1995) 309–317 [`arXiv:hep-th/9411178`].

J.H. Schwarz, "The Power of M-Theory", Phys. Lett. **B367** (1996) 97–103 [`arXiv:hep-th/9510086`].

J.H. Schwarz, "Lectures on Superstring and M-Theory Dualities", Nucl. Phys. Proc. Suppl. **B55** (1997) 1–32 [`arXiv:hep-th/9607201`].

C.A. Scrucca and M. Serone, "Anomalies and Inflow on D-Branes and O-Planes", Nucl. Phys. **B556** (1999) 197–221 [`arXiv:hep-th/9903145`].

A. Sen, "Strong-Weak Coupling Duality in Four-Dimensional String Theory", Int. J. Mod. Phys. **A9** (1994) 3707–3750 [`arXiv:hep-th/9402002`].

A. Strominger and C. Vafa, "Microscopic Origin of the Bekenstein–Hawking Entropy", Phys. Lett. **B379** (1996) 99–104 [`arXiv:hep-th/9601029`].

W. Taylor, "Matrix Theory: Matrix Quantum Mechanics as a Fundamental Theory", Rev. Mod. Phys. **73** (2001) 419–462 [`arXiv:hep-th/0101126`].

P.K. Townsend, "The Eleven-Dimensional Supermembrane Revisited", Phys. Lett. **B350** (1995) 184–187 [`arXiv:hep-th/9501068`].

A.A. Tseytlin, "On Nonabelian Generalization of Born–Infeld Action in String Theory", Nucl. Phys. **B501** (1997) 41–52 [`arXiv:hep-th/9701125`].

G. Veneziano, "Construction of a Crossing-Symmetric, Regge Behaved Amplitude for Linearly Rising Trajectories", Nuovo Cim. **A57** (1968) 190–197.

G. Veneziano, "A Stringy Nature Needs Just Two Constants", Europhys. Lett. **2** (1986) 199.

M.A. Virasoro, "Subsidiary Conditions and Ghosts in Dual Resonance Models", Phys. Rev. **D1** (1970) 2933–2936.

E. Witten, "String Theory Dynamics in Various Dimensions", Nucl. Phys. **B443** (1995) 85–126 [`arXiv:hep-th/9503124`].

E. Witten, "Bound States of Strings and p-Branes", Nucl. Phys. **B460** (1996) 335–350 [`hep-th/9510135`].

T. Yoneya, "Connection of Dual Models to Electrodynamics and Gravidynamics", Prog. Theor. Phys. **51** (1974) 1907–1920.

Index